PROFIT PRO
"The Art & Science of Menu Profit"

Bo Bryant & Thax Turner

ProfitProPlus.com

Copyright © 2015 Bo Bryant & Thax Turner
All rights reserved.
ISBN-13: 978-1511822589

ISBN-10: 1511822589

DEDICATION

We would like to dedicate this book to the scientists. In a world of art we often times lose sight of the science. To those fully immersed in the studies and data associated with human psychology, sociology, pattern recognition and neuro linguistic programming... thank you!

PROFIT PRO

CONTENTS

	ACKNOWLEDGMENTS	
1.	MENU MINDSET	1
2.	BAD ASS BRANDING	10
3.	PLANNING	15
4.	FOOD COST	25
5.	FORMATTING	44
6.	CONSUMER BEHAVIOR	54
7.	THE RULES	88
8.	MENU DEVELOPMENT	112
9.	BONUS IDEAS	133

PROFIT PRO

ACKNOWLEDGMENTS

I have met a lot of amazing people along the way in this portion of my restaurant journey. To my dear friends Tim Kirkland and Mike Koenigs for motivating me to get started. To the artists and scientists of this business that have pushed the boundaries and not been afraid to fail. To Troy Guard for pushing the boundaries of creativity and Leigh Sullivan for constantly branding and creating. To Mark Dym for impressing upon me how import it is to see a vision through, no matter what the so called experts say! To Neil Strauss for bringing the stories of "The Game" to all of us and for introducing me to NLP. To Dr. Lynn, Professor from Cornell; for amazing studies, information and tested theories (and some students who create controversy).

PROFIT PRO

MENU MINDSET

"Once your mindset changes, everything on the outside will change along with it."
— Steve Maraboli

I always start with mindset but this book really pushes me to question my form. In most consulting projects where I help people understand the value of the menu, I usually focus more on branding and the importance of form following function. While branding is critical, the imperative is still mindset. In order to get started properly we need to discuss what the menu really is. To most people the menu is a price sheet. To those who think that... I believe you would be wrong.

The menu is your biggest, most critical marketing asset. The menu is what people see when they look you up on line. It is what your potential guests search for when deciding to dine with you. It is what people use to decide where to eat when they stop in front of your restaurant without ever stepping inside. It is also the tool that makes word of mouth marketing positive or negative. The menu is a tool that will create confusion or compliment your brand. What's more, your menu is a money making tool that the majority of operators use improperly. So strap in, grab your highlighter, make notes, dog ear the pages and get ready to change the way you think about this most critical element of your business.

Let me start first by asking you a question? WHY? Why do we write menus the way that we do? Who said it is the best way? When you wrote your first menu, you likely followed an outline based on the predictable nature you have grown accustomed to when you eat out. Appetizers on the top left hand side of the pages, entrees on the right page. Whose rules are these? How do they know it is the best way? Why does no one ever question these rules?

RANDOM FACTOID:
Stupid rules exist everywhere. Take for example the "I before E" rule. While this rule applies to around 44 words there are about 923 exceptions to this rule.

The factoid above is merely an attempt to get you thinking about implied rules or social rules that exist because we have never thought to question or challenge them.

This book is going to address common menu rules and tons of other menu rules you were not even aware of. Before we get started I hope I can impress upon you the importance of questioning rules. To quote from my favorite movie, the Matrix...

"These rules are no different than the rules of a computer system, some of them can be bent, others can be broken."

By all means the rules should always be questioned at the very least!

The true mindset in understanding a menu is understanding two critical components. These components also happen to be the secret overall formula to all business. Are you ready for this?

The secret to a successful menu (or business) is the combined balance of art and science. Voila! It really is that simple in theory but the application in any business usually butts heads with a severe juxtaposition. This is because most businesses are run by a person or people with a dominant position to one side or the other. Rarely have I found an entrepreneur or partnership that has a balanced blend of these two positions (or mind sets). What's worse, there is no right answer in one position or another.

Here is an illustration I use to describe the two types.

Scientist:
The scientist is numbers minded, process driven and thinks that the fast way from point A to point B is

indeed a straight line. The scientist relies on data and analytics to reach conclusions.

A  **B**

Artist:
A: The artist is emotionally minded, relying on touch, feel, aesthetics and gut. The artist has an advantage in the starting point of a process because the data is raw, indescribable and thus usually starts a point further in front of point A but also usually ends a point further away from B. While the scientist looks for the fastest way, the artist looks for the best way. The best way is subjective of course and the best way is not always the right way. The same could be said about the fast way. The fast is not necessarily the best way either. The artist will almost never take the straight path but rather veer off course, spend more time in discovery and try multiple different ways to accomplish their goals. Look at this illustration as an example.

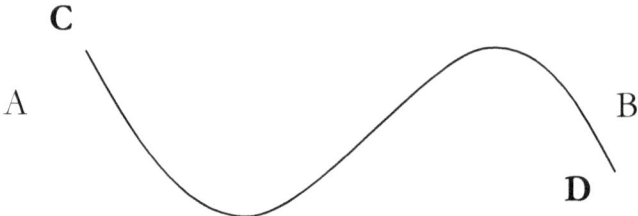

Again, neither approach is wrong nor are they linear.

They exist on separate planes and are often hard for one side to relate to the other.

Now I will reveal the true magic of this mindset. The scientist is practical and driven by data. Data translates into profit but pure data has no soul.

The artist is all soul but often times will miss an opportunity for profit. I know many a great chefs and owners who are artistic and make money in spite of themselves because they don't understand the science and a few scientists that make money but do not embody the soul or vibe a great concept can put out. In this industry I see far more artists than I do scientists. It would be fair to say that food is an art. That being said, it is most certainly fair to say that business is a science. If that is true, the optimum compliment for any business would be to achieve the balance of art and science. Now look below at the illustration when you combine these two mindsets.

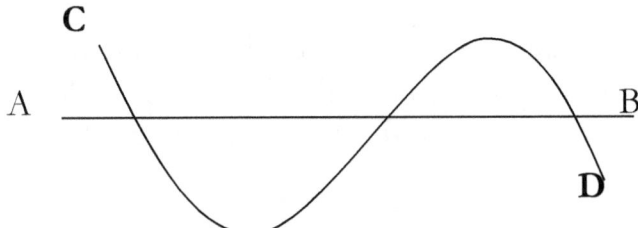

If you noticed the symbol... good job! If you didn't, try rotating the image 90 degrees and you should see something similar to this.

$

I draw my truest inspiration from this mindset and the importance of art and science. My hero and muse for the balance of these two notions come from Leonardo Da Vinci.

Da Vinci's art is most notably the Mona Lisa. His science not only brought about new ways to create paint and applications of his paint that represent in a near 3-d illusion, but his science also imagined things like the first drawing of the helicopter, the airplane and modern day weaponry.

Now think about your current menu and ask yourself what approach you have taken. Is your menu scientifically designed or artistically designed? Is it all form and no function or all function and no form? Could you achieve the Da Vinci code of balance? When you are finished with this book, you will!

In this book you will understand the importance of design and the importance of engineering. Basically you will get the tools needed to apply art and science to your menu which will result in

more profit and more brand alignment and likely more satisfied customers. If you are a big picture person, you will also walk away with an even greater understanding of how to apply art and science to everything you do.

Before diving into the menu, keep this in mind, a designer is no more an engineer that a doctor is a mechanic. My late father was a mechanic and I can hardly change my own oil. My stepfather is a doctor and I can hardly get a Band-Aid off of the paper without making it stick to itself. My point is, just because you hire a graphic artist to do your menu doesn't mean that they understand the engineering, same as if you were to hire a menu engineer and expect them to put out a well-designed graphic product.

There are a few other things to consider when looking at your menu. Here are a few questions for you to ponder.

What does your menu currently say to your guests?

- **Is your menu torn or dirty?**
- **Is your menu bland and boring?**
- **Is your menu hard to read?**
- **Is your menu a creative reflection of who and what you are?**

If you are not looking at your menu as the most important marketing tool in your business you are looking at it wrong.

We will cover a lot of the rules about the menu and what to do and what not to do at the conclusion of this book. The mindset part of the book is really about the way you think and moreover the way you should think if you are not in line with what I am telling you.

Your menu should be measured. You should know every single item on the menu and how well it sells. You should know your food cost on every single item on your menu to the very penny and percentage. Your menu should be easy to edit, reprint, adjust, add and delete any item at a moment's notice. We will cover a lot of that in formatting and design along with the importance of form and function.

Last but certainly not least, your menu needs to be a creative expression of the authenticity that represents both you and your brand. There are no rules on what a menu can be made of, for the most part. Get way outside the box and push your creative boundaries. If you are not creative, find people who are. Please understand, I am only talking about the creativity or design of the menu as a vehicle. The engineering is 100% on you or whomever you trust that has read this book and understands the engineering principles we are about to discuss.

BAD ASS BRANDING

"Brand is a noun. It is a verb. It may be about what we do. But, overall, it is all about what is in the mind – the mind of the consumer and the mind of the employee."
— *Sudio Sudarsan*

Branding is often times a very confusing term. Most people think the term is synonymous with marketing. This couldn't be further from the truth. Marketing is the way to deliver and measure a message and said messages impact on stimulating business, sales or involvement in a concept. I define branding as every part of the experience a customer has in your restaurant or concept. Branding is not just your logo or your concept's name. It is the menu, the way your employees look, the way your food tastes and represents the pricing strategy you use, the way you business looks from the outside and the inside. Branding goes on and on forever as it truly is every part of the experience your guest or customer has.

Branding as it applies to your menu needs to be line with the comments mentioned above. I find the most helpful way to help people understand a concept is through questions and answers. I also believe examples, imagery and storytelling also helps make the concepts tangible. So let's jump right in to how to brand your menu.

The first part of branding your menu starts long before the design work. These are the questions you need to ask yourself before you can begin.

I will get to these questions in just a moment but first you need to understand brand image and the importance of knowing who you are.
If you do not know who you are, no one else will.

In my last book, Whisper Marketing, I covered the very same subject. The story goes a little something like this.

Imagine you were at a cocktail party and standing across the room was your idol. This could be anyone from an actor to a Nobel Prize winner. It's your idol. Get that image in your head. Imagine the feeling of the butterflies you would have in your stomach. The way you might feel star struck. Now imagine that person is walking right towards you. It seems they have heard you have a restaurant and they love cooking. They are a self-professed "Foodie". Now your palms are sweating and your mouth has gone dry because that person is looking you down as they are two feet away with an expression on their face that is saying hi. And sure enough, your idol says. "Hi". "I hear you own a restaurant". "What kind of restaurant do you own"?

What would you say? Would you sound like an idiot? Would you talk them to death trying to throw everything at them but the kitchen sink? Would you confuse the shit out of them? Have you had this talk with anyone else before? If you have, you know what I am talking about. I too have been there. In my first restaurant I thought I could succeed by being everything to everyone. The big lesson I learned from my first restaurant to my second restaurant was that you cannot be everything to everyone.

You should seriously put this book down right now and see if you can script a quick two sentence way to describe your restaurant in a way that would be both impactful and make you sound different than every other restaurant out there. Because if you are not thinking how to make yourself sound different, then you will only appeal to those that live right in your general neighborhood and that is not the type of reach you want to limit yourself to.

The two sentence speech is also known as your elevator pitch. The elevator pitch is intended to be your hook. The quick and easy way to describe your concept in a way that qualifies or disqualifies the interested party. So what make you different? Even if you are the same, how can you make yourself sound different? There are some principles in anchoring where you may want to make you brand sound or look like another brand, but this is a dangerous and short lived application and one I would strongly suggest against.

All of the information I have just covered is often times rolled up into one simple term. The term I specifically use is called your "ASP" or your Authentic Selling Proposition. Your ASP is both your two sentence elevator speech and it should or could also be used in your tag line. Tag lines are very effective ways to capture a brand and impress your brand in people's minds. Here are some examples of an ASP used in a tagline. See if you can identify every one of them before I reveal the answers.

"Better Ingredients, Better Pizza."
"Melts in your mouth, not in your hand."
"I'm loving it!"
"Can you hear me now?"
"When it absolutely, positively has to be there on time."
"What can brown do for you?"
"31 Flavors"

If you guessed…
- **Papa Johns**
- **M&M's**
- **McDonald's**
- **Verizon**
- **FedEx**

- UPS
- Baskin Robbins

...you nailed it!

What is your brand?

What do you want people to think of when they hear your name?

What is your ASP? (An ASP is your authentic Sales Proposition)

It may seem like I am taking the long way around to get to the menu but understanding your brand and your brand's image, this information is critical.

You brand has to be Bad Ass if it is going to stand out and capture people's attention!

PLANNING

"By failing to prepare you are preparing to fail."
-Benjamin Franklin

I call this chapter **Piss Poor Planning** because let's face it, the majority of us avoid planning because it's work that doesn't appeal to us. Appealing or not, it is one of the most critical elements of a successful menu and hopefully I can make an argument that will appeal to you.

The number one rule I tell every culinary student I teach or any chef I work with is this...

"Just because you can, doesn't mean you should".

What I mean by this is quite simple, there are so many amazing recipes and food combinations out there that could and probably would enhance your menu but if the item doesn't fit, if the price doesn't fit and/or if it ends up making the menu too big, you have to make some decisions to do less than your total capability in order to achieve the optimum menu.

Let's talk about the critical planning elements and

how to execute them.

The first thing to do when writing a menu is to just let it flow. Write your ass off! Create as many unique and amazing dishes as you can. Then you will have to take a hatchet to the menu and chop it all up. This is called the editing process. Here is what to consider when editing your menu. Follow these simple rules and you will stand a better chance of making a menu that works and fits your brand.

NEW MENU RULES

- **Do it better than anyone else!**
- **Do it different than everyone else!**
- **Do not offer lip service!**
- **Do not try and be everything to everyone!**
- **Make sure each perishable item you use on the menu has at least 3 applications.**
- **Make sure the items you offer are in season and readily available.**
- **Make sure you know your food cost of each item and that the pricing fits your**

demographic.

- **Make sure that all items are in line with your overall pricing strategy.**

KNOW YOU REASON
This by far the most important part of your existence. Why should you exist? In an interview with the CEO of Whole Foods, when agreeing to have a mentorship meeting, the first question he asked about the business why "Why should you exist?".

This was one of the hardest yet greatest questions I had ever heard.

Why should you exist?

- o What is your reason to take up space in the landscape out there amongst all the other businesses in your competitive space?

- o What do you offer that is better, special, different?

- o Why do you deserve to be here?

- o What will you give people that they cannot already get?

Think about these questions. They might seem silly or even meaningless but I assure you, if you can answer and execute against these questions you have a much higher likelihood of succeeding.

The answers to these questions need to become your "guide rails" or the rules to framing your business.

When it comes down to your menu, ask yourself if the items on the menu satisfy these questions. If they do not, you should consider working them in a way that they will.

KNOW YOUR MARKET!

Who is your audience?

RESIDENTIAL
The best way to find out who your audience is, is by doing some local demographics research. The best place I have found to do demographic research is on a free website called **zipskinny.com**

On this website you can enter in your zip code and a report will populate telling you all of the demographic information on the residents that live in that particular zip code. This is great information for one part of your audience, which is residential. This report will give the age, education, income, ethnicity, percentage of single vs. married households, what they do for a living and much more. You can also compare zip codes where you are located to other zip codes around you. I use this site religiously when studying my locations and when looking for new locations. Keep in mind if you are looking for a new location you would be wise to compare your current demographics to your potential next site to make sure they are comparable.

BUSINESS

You can find out what your local business demographic is by contacting your local chamber of commerce or Better Business Bureau. Most of these organizations have local business demographic information in reports to help you identify your market. Most restaurants have at least two day parts of service and most of these services will appeal to at least two separate demographics. If you are in a high traffic shopping or tourist district you could have a few more demographics to analyze. Another site I like to use for business analytics is Google Plus Local. When searching GPL you can enter in your zip code and it will show you a map of your local business in a consolidated area.

NOTE: The target demographic areas for most restaurants are located within a 3 mile radius. Dining studies have shown that majority of workers and residents do not travel much outside of a 3 mile radius to eat.

KNOW YOUR COMPETITION!

Who is your competition?

Again, keeping in mind that most people eat within a 3 mile radius of work and home, you are best served to travel the 3 mile circle around your business to see what other restaurants are in your neighborhood or business district. When doing so you should do as much research on these concepts as you can. Here are the research steps I take when looking at my competition.

LOOK THEM UP ON LINE

Review their menu:
I like to look up the menu of each and every competitor in my area. I print the menus and look for a few common threads. What are their price points? Is there a local item all of them serve? Who is similar to me and how will I create a point of difference in my market place if I do have a similar competitor?

READ THEIR REVIEWS

Read their Yelp reviews. Read City Search, Travelocity and any other review sites that are popular in your area. What I am looking for in this review study is twofold. I am looking to see how sophisticated the audience is and what common thread they are all talking about. Is this a heavily service oriented audience? Is this a very value driven audience? Are there a lot of complaints about pricing? This common thread will help you be better than your competition because you will be looking at the sum of all parts and you will be better able to put your finger directly on the pulse of the audience you are serving.

BE A CUSTOMER

There is no secret to being a competitor and a customer. This is not a cutthroat business. It may seem that this business is cutthroat but if you follow the path I have laid out thus far, it would be counter intuitive to think any other way. If you are trying to clone an existing business and you are ashamed of that, you will likely be in the enemy mentality, but if you are looking to be the best and do things better than anyone else, you will likely have no enemies and therefore you should welcome friendly competition and collaboration. I know I am going off on a diatribe here but I can't emphasize the importance of community enough. You have to have the right mindset and that includes understanding the importance of community. You either are or will be joining a community and collaboration is proven to be a better friend than an enemy. I don't see the independent restaurants in competition with each other. I have always believed that the independents need to stick together to illustrate the importance of local, neighborhood concepts. As an independent I always looked at the big chain stores as my direct competition and at my fellow independents as my brotherhood. When I open a new restaurant I will go around to those concepts and introduce myself to the owners, managers and staff. I always let them know who we are and what we do. This is a great audience to practice my ASP and two sentence pitch on. I also make sure to tell them that if I can ever do anything to help them out that I will. If they ever need to borrow anything I am there for them. Every time I have approached a community like this I always end up having allies in my efforts to succeed. I could tell you countless stories of staff from the other restaurants that have come to my place to eat at a discount, drink until the wee hours of the morning and even eventually come to work for me when their current employment has run its course.

When I do become a customer and I do become allies with my competition, I end up learning a lot about the area that no report or analytics study could ever tell me. I learn who the players are. Where the talent is. What some of the unspoken rules are. Who to watch out for. Who not to hire and so on. I don't always take everything at face value but any information is worth considering, especially if it was information I didn't already have. I also find it very helpful in learning things about the community and when big events happen, what has worked for others and what hasn't, when the slow season is… the list goes on and on. I hope you can see the importance of this mindset and strongly urge you to adopt it. You never know when someone around the corner will save your ass!

This may also seem counterintuitive but I never shit where I sleep. I do not go out trying to recruit people in my 3 mile circle. These are my people, my allies and my neighbors. That is not to say that I won't travel outside of my circle to find great people but you do not want to be that guy or gal stealing talent from your neighbors. That will get people talking about you in a bad way. If you kill your local neighbors with kindness they are bound to reciprocate. I will often recommend people to another restaurant in my neighborhood that is different from my concept and often times my neighbors do the same for me. It keeps the money and the people in our local area and the consumer public appreciate a vibe and feel of community and there is no marketing effort of concept alone that can trump that type of feeling.

Okay, off the soap box and back to the menu!

Another critical practice in planning is R&D (research and development). I approach R&D in many different ways but at the end I find the most important part about writing a menu and creating amazing food is that I am not the judge of it. The people are the judge and you must go to the people. If you have a current restaurant and you are working on new menu ideas, a great place to start is with the "Daily Special" or the "Feature of the Day". You need to track the popularity of your specials so you know what is working and what is not. Many times I will use specials as a sneak preview of what is to come on the new menu. If you are a new concept it is good to bring in a focus group similar to the audience you will be serving to get their feedback on your items. These people are called focus groups and often times you will get better and more honest feedback than you ever would from friends and family.

FOOD COST

"What gets measured (and clearly defined) does get done."
— *Mike Schmoker*

I do apologize for what I am about to do. If you are like most restaurant people this part is painful. For that I am sorry! I have not established any true authority to you so far but I cannot impress upon you how important this chapter is! IT IS THE MOST IMPORTANT CHAPTER OF THIS BOOK!

This really should have been the first chapter because this is the place where more mistakes are made in the business than anywhere else. Food cost has become a passion of mine for all of the wrong reasons. I am a very creative person by nature and as most creative types I tend to stay away from numbers. Quite honestly they put me to sleep. But then I found the difference that knowing my numbers made both financially and professionally. Getting other people excited about numbers has never been easy. However, in my extensive years of consulting I discovered quickly that most concepts didn't need another cook in the kitchen or another creative person added to their fold. Where most operators struggle is with the numbers. Justifiably so because… we are creative types by nature and that is likely why we are in this line of work.

In most concepts the balance of every dollar taken in as revenue commits 60% that is taken out in the combination of food cost and labor. It is with this number in mind that hopefully I can appeal to your sense of logic and understanding when it comes to the importance of food cost. Food cost has two very critical components that create one basic equation. The two elements of food cost are these...

THEORETICAL FOOD COST & TRUE FOOD COST

Theoretical food cost is what your food cost should be based upon your recipes, portions and mix of product sold. I will show you an illustration of how this works momentarily. Basically put, if you could measure perfection and aggregate that perfection based upon what you sold, the output number form this would be your **"Theoretical Food Cost"**.

In contrast, when you do your inventory at the end of the month and you use this basic equation to follow, you will come up with your **"True Food Cost"**.

True Food Cost =
(Beginning Inventory − Ending Inventory + Monthly Purchases/Sales.)

You may be asking what the difference is and that is okay, most people do. The reason for knowing both of these numbers is so that you will be able to effectively measure your deficiency gap.

The deficiency gap works like this. When you know what your food cost should be and then you do your end of month inventory and the equation to figure out your true food cost, you will likely find a gap. In studies of the best concepts the appropriate gap between these two numbers should be no more than 3%. However, when I consult with clients that do not have these practices in place and I implement them…

We find the gap on average to be about 9%.

When not using a Food Costing Program or software you are leaving a ton of money on the table. Can you imagine what that 6% of your total revenue would mean to your bottom line if the gap was closed and that money was in your pocket? This is why food cost and understanding the importance of both your "True and Theoretical" cost is so critical to your menu.

There are many food cost programs out there. Some are free, some are cheap and others can cost hundreds of dollars. Most food service providers, especially the broad-line distributors, have these programs complimentary for their clients. If you cannot find one just let me know and I will send you the one that I use for free.

Let's get in to a little more food cost theory so that you understand how to manage your recipes and measure them properly.

The basic part of any good food cost programs starts with your inventory. I look at inventory two ways. You have ready manufactured products that you buy and ready manufactured products that you make.

Here is an example of the two. You could buy salsa and that would be a ready manufactured product. You could buy the ingredients for salsa and make a batch from your own recipe and that would then become a ready manufactured product that you made. It's that simple. You are either buying something made or buying the ingredients so that you can make it. Before you can start doing food cost you have to have all of these items broken down to a portion cost. I call this part of the food costing formula the master product list.

Once you have a master product list you will need to calculate your cost for every item on that list. I use a rule of breaking down my cost of every item to the lowest common denominator. For everything I do I measure by the ounce or by the each. By the each would be something like a proportioned steak, for example. That is how I can get one universal cost whether I am counting a batch recipe by the pound of a plate recipe item by the ounce.

Here's where it gets tricky for some. When it comes to manufacturing your own product, figuring out the cost is quite a bit harder. Here are the steps you need to consider when creating your own manufactured items or batches. First, understand that everything can and should be measured. Measuring creates standards and gives way to proper portion control. Improper portion control is the leading cause of the large gap between "Theoretical" and "True" food cost.

Let me show you the salsa recipe as an example and each ingredient at a time.

Here's my recipe:

SALSA
5 lb. diced Roma tomatoes
8 oz. Diced Onion
1 oz. chopped cilantro
1 oz. pureed chipotle in adobo.
1 oz. Salt
1 oz. Cumin
1 oz. Minced Garlic

Now here is the breakdown of my ingredients.

Tomatoes – 20 lb. case fresh Roma tomatoes = $20 case
After I core and chop 20 lbs. of tomatoes I get 19 usable lbs. This is called yield. The yield weight is all you can count in your food costing.
To factor for yield I would multiple 19 lbs. x 16 oz. and / the total by $20
Which would = $.0658 per usable ounce. Now I would multiply my oz. cost by my total ounces called for in the recipe. The recipe calls for 5 lbs. or 80 ounces. My total cost for 80 usable ounces is $5.26. Do you see how the yield factors? If I had just used lbs. from my 20 lb. $20 case, I would have come up with a cost of $5. This would be inaccurate by over 5%. Let's assume that all of the other items I put in this recipe are already pre-done in the form they are called for in the recipe. Let's look at the cost of my batch.

5 lb. diced Roma tomatoes	$5.00
8 oz. Diced Onion	$0.48
1 oz. chopped cilantro	$0.52
1 oz. pureed chipotle in adobo	$0.42
1 oz. Salt	$0.04
1/2 oz. Cumin	$0.82
1 oz. Minced Garlic	$0.56
Total for the batch	$7.84
Total Weight of batch	92 oz.
Total Cost per Ounce	$0.085

Once I know my batch cost of something I can figure out my plate cost or menu cost of the item using that product.

You would have to repeat this process on every item you make.

The easiest way to figure out a yield on any item is to start with the cost of each individual item, reduce the product to total usable product and divide that usable amount of product by the case cost.

Note: This should not be done by hand. Many times people will do their food costing by hand or even on an excel worksheet but unless you use a program that will allow you to update your pricing monthly, you will never have an accurate cost.

Profit Pro Plus will allow you to import your master product list and pricing and will even be compatible with your vendors so that they can send you updated pricing that can be uploaded in your program.

Once you have the cost for all of your batch items and menu items you can now start to figure out how you should price your items.

Here is an example of a recipe that we have the cost for.

Item	Measure	Units	Unit Cost	Total	% of Cost
Bun	Each	1	$0.60	$0.60	21%
Burger	Each	1	$1.40	$1.40	48%
Lettuce	Oz.	.5	$0.08	$0.04	1%
Cheddar	Each	1	$0.24	$0.24	8%
Tomato	Oz.	1	$0.06	$0.06	2%
Onion	Oz.	.5	$0.08	$0.04	1%
Sauce	Oz.	1	$0.16	$0.16	6%
Fries	Oz.	6	$0.06	$0.36	12%
			TOTAL	$2.90	100%
			SELL	$9.99	
			FOOD COST	29%	

I have filled in the sell price for this example.

Considering this example, I would like you to review this case study for a better illustration of how the best in the business go about impacting there profit in a responsible way.

Here is a Menu Cost Strategy case study that will change the way you look at impacting your profitability.

CASE STUDY

South Mouth – Memphis Style Hot Wings
1650 Broadway
Boulder, CO 80235
March 2015

There are 6 different strategies to impacting your Menu Cost and 1 Best Practice!

- Portion Adjustment
- Product Adjustment
- Price Adjustment
- Menu Adjustment
- Deflection
- Elimination

*Best Practice – Use a blend strategy for best results.

CHALLENGE: Each year the price on the cost of the Bag in the Box or "BIB" Soda product and the packaging increases. As the product cost increases each year it forces operators to either change their pricing strategy or lose gross profit. This is true for just about any product that a restaurant buys. This is a simple example of how we can impact and measure our product so that we can manage our profit according to the best practices in the industry.

In this study we looked at the impact of a QSR (Quick Service Restaurant concept called South Mouth) that sells a single size, 20 oz. fountain soda.

The restaurant was paying around $68 for a 5 gal "BIB" of soda. The ratio mix is 3 parts water, to one part "BIB" syrup. This equals a yield of 2,560 ounces per BIB which equals .027 cents per ounce.

We needed to know how many ounces of product we served over a 28 day period and how many ounce we bought over a 28 day period.

Why did we do this, because we wanted to know the "True Cost". In this case study we wanted precise data and since we have free refills knowing how much soda it took to fill one cup just would suffice. But we wanted to identify how much soda went into a 20 oz. cup with ice.

Here is what we found.
However, after adjusting the price of our soda to reflect what we found in our true cost our 14 oz. turned into 19 oz. We consider this fair due to refills, mistakes, over pours etc. That increased our

In a 20 oz. plastic cup (.24¢)
we used 4 oz. of ice (not factored in price)
with a lid (.09¢)
we averaged 19 oz. of soda per soda sold (.51¢)
and included a straw (.005¢)
Giving us a cost of = 85¢
Or a cost of = 45%

This is not acceptable when the industry standard tells us we should be at a soft beverage cost of 35% or less.

This means we need to have our cost down to .70¢ based on our $2.00 menu price to hit 35% (This was a contract cost and we couldn't get it lowered so we had to consider a price increase.

A menu price increase would need to be $2.50 in order to get to a 35% beverage cost. That was a 25% price increase and before we just reacted we needed some more data.

In a study of our competition we found that the average price in our concept type in the local area was $2.35.

We did not want to be the highest but we did not want to have such a high soft beverage cost.

We needed to consider other options.

We asked ourselves the following 6 questions in order to devise our strategy.

- **Can we adjust our Product?**
- **Can we adjust our Price?**
- **Can we adjust our Menu?**
- **Can we use a Deflection Strategy?**
- **Do we need to Elimination the item?**

Best Practice – (Use a blended strategy for best results)

Q: Could we make a portion adjustment? If so, how could we do it?
A: We could move to a 16 oz. cup.
   Conclusion – *The guest perception would be unfavorable as we would have to lease the price the same to capture the % we needed.*
A: we could move to a competitors product.
   Conclusion – *The price per ounce was almost the exact same and the switch was not worth the hassle or the guest push back.*
A: We could do a cost comparison on canned soda and possible switch.

Conclusion: *Very favorable. The can of soda cost .65¢, we could sell for $1.99 and have a cost of 33%. It would also free up space not having to use cups, the bulky fountain machine, the ice, the lid, etc. Unknown: What the guest perception would be.*

A: We could move to an ice that would fill up more surface area thus causing us to use less beverage per 20 oz. cup

Conclusion: Pellet ice is what we found as the best ice to cover the most surface area in a cup thus allowing us to go from 19 oz. per soda sold to 14 oz. which reduced our overall cost to .73¢ getting us close to our ideal cost.

FINAL CONCLUSION:

Product Strategy: We adjusted our ice machine to go from traditional cube ice to the pellet ice. (Subsequently, we received no pushback but rather many comments from guest who enjoyed chewing on the ice). This reduced our oz. per cup average down 4oz.

Price Strategy: Additionally we did make a price adjustment. We knew the market would bear us charging $2.25 as the average competitor was at $2.35. It allowed us to continue our value proposition while still being more than competitive.

Menu Strategy: We did not make a menu adjustment on the soda outside of price but we did use a Deflection strategy.

Deflection Strategy: We added 3 flavors of handcrafted lemonades: Strawberry, Blueberry and Ruby Red lemonade. These drinks did not have a free refill option and due to their uniqueness commanded a higher price at a cost that held more profitability than our fountain soda. In our deflection strategy we merchandise sales and marketing material around the cash register along with a recommendation upsell technique where the cashier would no longer ask if they wanted a drink but instead if they would like to try a handcrafted blueberry, strawberry or ruby red lemonade. Beyond it being a nice compliment to our category it also gave us some additional point of differentiation.

- **Product Adjustment - Yes**
- **Price Adjustment - Yes**
- **Menu Adjustment - Yes**
- **Deflection Strategy - Yes**
- **Elimination - No**

*****Best Practice** – (Use a blended strategy for best results) **YES!**

A quick rant about trying to save money on purchasing: You could go to the local food warehouse and buy the product yourself to try to save a little money on your ingredients but at what cost? Would you be saving any real money to get your own vehicle, waste your time and gas to become your own distributor to save a few pennies? Let me not beat around the bush. If you are running to the local food warehouse to buy your food to save some money, you are doing it wrong. I will explain why but that is another book. Just know this; no successful, sustainable restaurant brand that is of any substantial size or quality would ever do this. You cannot manage a lower overhead than a distributor and you if can you are doing it wrong likely compromising someone's safety or the safety of the food you are preparing. If I am wrong and you have uncovered some magical formula then you should close your restaurant and start up a distribution business.

By working this exercise when building your menu you will have the exact science of what is doable and what is not. With all of that said you should dismiss the idea of having a target food cost item for every single item on your menu. Just like the grocery store sells milk at a very thin margin to get people in the door, because they know they milk price is important but you will buy other things that you likely have no idea what fair market value for those items are. Restaurants can do the same thing on the menu. One of the reasons I say you should do everything on your menu better, different and in a unique way is so that your customers can not compare your items and their pricing to the "Fair Market Value" of a like item. When consumers use "Fair Market Value" when judging your pricing you will lose. The only way to let them use this tactic is to have things like everyone else. I will discuss this tactics more in layout and rules. But for now let's get back to food cost. Just like the grocery stores, you can still make plenty of money even with items we call "loss leaders" like milk. The individual items food cost is not as important as the compressed value of all of the items you sell and all of the items combined food cost.

I will show you an example momentarily but first let me point out the importance of research. The typical spaghetti dish in my neighborhood sells for $18. The typical cost for an item like this is 20%. Notice the restaurants didn't mark down the price to hit a higher food cost closer to their ideal. With items that have a higher perceived value than they actually cost, the fair market value is actually inverted. You have to take advantage of opportunities like this to offset the challenges of high food cost items. That is why it is so important to know what your competition is charging. This way you can take advantage of items you may have otherwise priced lower.

Here I am going to show you an example of how the blended food cost means more than the individual items food cost. Again, most good food costing programs will have a report like this built into the software.

Item	# Sold	Cost %	GP$	Sell Price
Burger	120	29%	$7.10	$10.00
Fish n Chips	100	35%	$6.50	$10.00
Chicken Caesar	88	26%	$7.40	$10.00
Steak Sandwich	140	40%	$9.00	$15.00
Ice Cream Sundae	60	22%	$4.10	$5.00
TOTALS	**508**	**33%**	**$3,647.20**	**$5,480.00**

This is an abbreviated report (in a number of ways) meant to simplify the example. As you can see in the chart above, even with a very popular item at 40% food cost and another at 35%, the compressed value of the entire theoretical food cost is only 33%. Why this example may seem high as an average the myth of a 30% food cost is a misnomer. There are always mitigating factors like labor cost in your market, rent, competition, etc. Don't use this 33% as a good or bad example. Just notice that with some items that have a high food cost the average is considerably lower.

Another point to illustrate is the Steak Sandwich. This example points out that while we often focus on food cost percentage we do so while missing the most important number. That is the Gross Profit Dollars. Even though this item has a 40% food cost it makes us over $1.50 more in revenue than the next closest item which has a much better food cost. Looking at this example; which would you rather have in your pocket at the end of the day: $7.40 or $9.00?

I know you are drinking from a fire hose right now and this may be a lot to take in. You may also be thinking that it shouldn't be this hard to write a menu. I am sorry to inform you but it is this hard. This is the science behind profit and without profit you will not make it in this business.

FORMATTING

*"If you don't know where you are going,
You'll end up someplace else."*
— *Yogi Berra*

Now we are getting into the fun stuff. Formatting is the section where we will look at your layout, delivery, cover and pages. These can be either conventional or unconventional. The basic rule for a menu is that while it may cost you quite a bit of money, you can decide when you will spend the money. You can either spend a lot up front for a nice cover where all you have to do is reprint black and white pages maybe on a cream colored paper or maybe you will spend a little bit on a printed paper menu that will require reprinting often. There are of course options in between but what I want to focus on is more about what not to do rather what to do. I will cover both but let's start back at mentality and practicality.

Let's start with some basic rules for formatting a successful menu.

- **Form should follow function!**

The menu has to be easy to change. There are always times when you will need to raise a price or take an item off the menu without doing an entire menu change.

Your menu should also be formatted in a computer program that you can readily access and edit. I prefer to do all of my menus in Microsoft Word. That does not mean that I have to compromise graphics or artistry. If I am doing a menu that requires background art or imagery I will have a graphic designer create my art work in Adobe and load the art in Microsoft Word as an image set to the background. I will then create text boxes over the art work so that I can edit my text any time.

Buyer Beware: *In the industry there are many companies that offer to do free menus for restaurants. I have seen it range from advertising companies to food and liquor purveyors. Here is the problem. The free menu is anything but free in most cases. It is a trap. When you agree to allow a service provider or a distributor do your menu, they will usually do nice work and usually at low or no cost or in exchange for doing business with them. Sounds great right? Wrong! The common tactic in this ploy is that the company will hold you hostage with your menu. They own the menu and the rights to it. If you were to leave the company they would maintain possession of your menu and you would have to start all over. The other challenge to letting someone else own your menu is the convenience of change. In this case, when you want to make a change, you will have to wait and depending on how many people that work on menus and how many customers want changes, you could be waiting quite some time. There are some reputable companies out there that will do all the work for you and give a working editable copy when they are complete. Just make sure you ask whatever company is offering this service to you if they give you the file and if it is editable when they do. If the answer is no, you should pass on this wolf in sheep's clothing.*

You should change your menu at least 2 times per year. Ideal would be 4 times per year.

As the seasons change so does product availability, product pricing, product availability and even our appetites. I believe that a well-rounded restaurant should change their menu 4 times per year. The rules I use are right in line with the seasons. I have a summer menu, a fall menu, winter and spring. During the beginning of each season, produce and proteins prices can take major swings. For example, middle meats such as tenderloin and ribeye reach their peak in late fall and early winter due to the demand over the holidays. In the spring pricing on lettuces and avocados go way down and the quality goes up. Also in spring and summer many new produce items come available that are not usually in season. There are also winter crops that only come available that time of year as well. Remember availability mean supply which equals low demand. Low demand = good price. The inverse is true as well. As human beings we also have prefrontal cortex triggers that go back to the dawn of our evolution. Triggers in the brain that tell us that as rule of survival, when it gets cold we eat more to put on fat for storage and lean hunting/gathering season and the same triggers kick in when it gets hot out that tells our brain to eat less and stay lean as food is abundant and readily available. Translated to modern mentality… eat less in the summer so you look good in your swimsuit. So remember: Winter = heavy & hearty while Summer = light & healthy.

Fine dining concepts will often take these rules to a whole other level. Many fine dining concepts will make menu changes monthly, weekly or even daily but that is not really the mainstream.

<u>Interesting Factoid:</u> In Japanese culture there are 12 seasons that revolve around crop cycles and different harvests. In many traditional Japanese homes they celebrate the food of the season buy changing their internal décor to match the season and the harvest. This can include everything from the color and design of plate ware to tapestry and home accents. The culture will celebrate most of the meals from that season just around the harvest crop.

- **Type rules according to Font!**

This is a pretty simple formatting category but important nonetheless. Font should not be sized smaller than 12 points. Smaller than that and it is too hard to read. There is also a rule of confusion discovered in advertising studies. The rule simply states that there should be no more than 3 fonts used in one application. More than three fonts create confusion. Fonts should also be easy to read. Stay away from cursive fonts and loopy or squiggly fonts in the body of your menu. You can get away with some of this in the category headers but even then you can create confusion. I'm not suggesting you be boring just make sure it is legible and doesn't take a lot of brain power. The last thing you want to do is exhaust your guest or make them frustrated.

- **Color Rules.**

No kidding! There is actually a study of the psychological affect color has on people. Especially as it pertains to food. Using the color blue on a menu is an appetite suppressant. Scientists have concluded the reason why blue is an appetite inhibitor is because it is the only color that does not naturally occur in food. At least not in a good way. While I love bleu cheese, the blue in bleu cheese is from mold and the human mind has a hardwired trigger embedded from our ancestors that eating blue is bad or poisonous. For those of you thinking "Hey, what about blueberries"? You would be right if they were actually blue but they are purple.

When you have time, Google the psychology of color and what it means. You will find out some pretty interesting things about the personality of people when you know these colors. You will also find some cool ways to market your menu to a feeling and this is a very powerful tool. I have listed some colors below as an example.

COLOR PSYCHOLOGY KEY:

EMOTION
Orange: Universal buying color, stimulates emotion & appetite.
Bright Red: Stimulates action, energy, excitement.
Burgundy: Richness, Warmth, Elegance.
Yellow: Happiness, Laughter, Hunger.
Green: Natural, Health, Tranquility.
White: Sterile, Pure, Honesty.
Purple: Wealth, Sophistication, Exotic.

TONES
Bright Colors: Excitement, Quick Pace, Energy, Stimulation.

Earth Tones: Calm, Soothing, Comfortable, Slower Pace.

These are just a few examples of color and how you can apply them to elicit the response that best suits the audience and emotions/actions you are trying to achieve.

-

-
-
- **Have fewer items.**

Note: Consumer confidence begins to erode when you have over 40 items on the menu. The thought is that if you do too much you cannot do it well.

To illustrate this point I usually ask my clients to think about their favorite restaurant. Imagine yourself at your favorite restaurant. Now think about their menu. How big is it? What are they known for? Can you sum up the brand in two sentences? If you are like most people you can. If you are like most people, your favorite restaurant has a simple menu. See where I am going with this? If most people's favorite restaurant has a small menu and you want your restaurant to be a favorite to most people then you should have a small menu.

- **Have more categories**
  - In a Stanford study on choice, it concluded that people are more inclined to make a decision when given less choice. In a separate study consumers also found that a menu with more categories created a perception of more value.

    (Source: Sheena Iyengar, Stanford University)

Here is an example of creating more categories without necessarily creating more items.

In a category like **APPETIZERS** you could have 10 items. Could you break that in to two categories and have 5 and 5? Could you break your items into two categories like this?

SNACKS & THINGS TO SHARE
or
SMALL PLATES & SHARE PLATES

Likely you can. And by creating two categories from the one you will achieve more eye traffic. I will get to the importance of that a bit later.

The same could be done with entrees. If you had an entrée category you could likely break that one category into three. Like this…

4 LEGS (beef, lamb, pork)
2 LEGS (Chicken, Duck & Turkey)
NO LEGS (Fish)
And for fun: **GREAT LEGS** (Wine)

Or you could do…
LAND (beef, lamb pork, etc.)
AIR (birds, fowl, etc.)
SEA (fish, caviar, etc.)

-
- **PRICING**

From a formatting standpoint there are some serious rules as it applies to pricing. This section is just about formatting so I will get into more details about the overall impact of pricing and the logic in the section titled **"RULES"**. Pricing as it pertains to formatting pricing should be approached from a commonsense standpoint. Pricing follows a funneling rule. The rule states that the menu should be written in an order of importance. Not the importance from the guest's viewpoint but of the business's viewpoint. We want the guest to make sound decisions based on selling principles. Sales work like this. We buy with emotion and justify with logic. That being said you should never put price anywhere but at the end. Price is a logic detail. The item's name and the description are where you capture emotion. This means the menu should be structured accordingly.

Item, description and then price. What is it? What's in it? How much does it cost? This will give you a chance to win them over with two out of three of your efforts which will help stack the favorable decision in your favor. Again, this is just the tip of the iceberg. We will cover a lot more on pricing in "RULES".

Now that we have covered the 7 important formatting positions, let's look at some practical application rules that cover the good, the bad and the ugly.

BAD

The number one "No-No" rule in menu formatting is laminating. Laminating creates an oxymoron in the menu world it is the most expensive cheapest looking thing you can do and it is impractical is hell. When you laminate a menu you are bound to have to redo the entire menu even if you only need to change one thing.

CONSUMER BEHAVIOR

"Science does not know its debt to imagination."
* *Ralph Waldo Emerson*

NLP, or neuro linguistic programming, is a form of behavior modification used in the professional world as well as in parenting. NLP involves an in-depth understanding of mental, cognitive and strategic thinking that are the processes behind how we behave. It is a very powerful skill for communicating. A simple gesture by a person toward us will be interpreted differently based on our past experiences, both good and bad. The action/reactions, our opinions, biases and values system are all stored and consulted. NLP strives to replace your negative responses with positive responses by learning the healthy patterns and behaviors of successful people that can enrich your life. – *Definition provided by: Maryanne Lane LMT*

Maybe you have heard of NLP and maybe the idea is new to you. By definition there are ways to train, un-train or retrain people in relationship to what they think about any given person, place or thing. This form of psychology is very applicable to restaurants and menus specifically. There are a number of preconceived notions that consumers have relative to restaurants and restaurant menus.

I like to play a little game to explain how NLP works and how we can extract the information we know to either gain an advantage or change a perception.

For this exercise I will need you to use your imagination. See yourself stopped at a stoplight. The last stoplight you have on your route from your home to your work. Imagine you are either late or better yet, you are distracted and not paying attention… you are zoning out! Can you see yourself in this situation? Can you recall a situation like this where you have been zoned out in front of a familiar stoplight and your foot has eased off of the brake and gone to the gas (subconsciously) the moment right before the light changes? If you are like most people you have done this exact thing at that exact stop light. Sometimes, you will snap out of your zone and think "Wow, I must be psychic or something". Sound familiar? Well, likely, you are not psychic. You have simply been given a glimpse of how your subconscious works. This event happens in the pre-frontal cortex of the brain. This is where pattern recognition is stored.

There are so many examples of this happening to you on a daily basis that if you had you apply real thought to these things you would never make it out the door in the morning. Your mind is wired to record, recall and relive these moments through receptors in your brain that have deemed these tasks at non-threatening or non-important. Basically, your mind has a hard drive with prerecorded tasked hardwired into your system. You mind also has a clock that can tell time and measure timing much better than you could ever think. While this little rant on neuro psychology may not seem important or relevant to the menu, studies I am about to share with you will absolutely change your mind about that. I will show you how some of these subconscious decisions, calculations and reactions can be used to your advantage and you will also be able to identify with them when you read them or at the least the very next time you see a menu.

So, without further ado… let's light this firecracker.

The first step into NLP as it relates to the psychology of the menu is how the menu is written vs. how the menu is read.

In a study by William Doerfler in the 1970s, he surmised that the menu had a sweet spot or sweet spots that people focused on when reading a menu. From his study he came up with what is popularly known now as the "focal map" or the Doerfler menu model which I will show below.

In contrasting form, researcher Sybil Yang from SF State had test subjects wear retinal eye scanning glasses to test where the eyes focused and in what order. Her hypothesis was that people read the menu much like a book from top to bottom and left to right.

This has created some controversy so I decided to work on this study independently in a properly controlled test environment.

Before I go into great depth on my challenge of this particular Sybil Yang study, it is worth mentioning that overall Ms. Yang's body of work is pretty impressive and she has been an amazing contributor to the science.

The Yang study was comprised of mostly college students and faculty with a group of 26. This group was in a laboratory looking at a menu on a 2.5' x 2.5' monitor with retinal glasses attached to a headband. The biggest flaws in this study were as follows.

1.) The study was conducted in a laboratory to a group conformed to classroom style learning (meaning open your book and read from left to right, top to bottom).

2.) The menu was set up in a non-traditional way which I believe threw people off. More about that as we continue.

3.) The menu itself was a monitor not a tangible menu that you can manipulate.

In the real world fraught with social pressure, time constraints and expectations, not to mention being at the whim of the service staff that checks back intermittently (so you had better stop talking and make up your mind), these conditions were absolutely missed. Also, I didn't agree with the Yang study because I have never seen a menu digested this way before I began my study.

While I am not a scientist by traditional standards I understand the value of a controlled environment, measuring and variables.

The outcome of my study did review a few things in line with the Doerfler study. People (in restaurants) do seem to follow an eye pattern path when reviewing a (physical) menu. Perhaps you will see in my conclusion that this sounds similar to how you read a menu when you are out to eat.

Before I could start the study I was looking for a confirmation. In a study of over 100 participants, the average person spent about 109 seconds reading a menu. Knowing that just over a minute and a half was not enough time for people to fully read and comprehend an entire menu. With this, I knew I had a basis for my study. Basically the evidence of the time study showed that people must scan menus. If they are scanning menus it might also stand to reason that they may have higher concentration of particular focus areas. What I found in addition to this example was that people read menus differently based upon the menu layout, format and number of pages. Next I arrived at the same conclusion reached by another study that stated people's eyes travel where relevant information exists. *(Noton & Stark,1971; Fisher, et al, 1981).*

What is meant by relevant information goes a little something like this. If I were to ask you to imagine a menu that was formatted like a book. The cover on the outside and two pages on the inside. Imagine you were to open the menu and see the two pages. One on the right and one on the left. Now imagine I was to ask you what was on the top left hand side of the page?

If you are like most people your answer would have been appetizers.

Now imagine in that same menu that you are looking at I was to ask you what was below the appetizers, you would likely answer soup or salad or a combination of both. Now imagine I was to ask you what was on the right hand page? Again, if you are like most people, you would say entrees. You may even go a little deeper into explaining some categories. If I took this exercise one step further and asked you what was on the back, you might probably say drinks, desserts, kids menu, restaurant information, the restaurant history or story. Most of these answers are pretty predictable. And that is where the story of how to engineer a menu starts.

It is with this information in hand that most customers receive a menu. The first thing a customer does in reading a menu is to do a quick scan to insure the "Relevant Information" is formatted as they expected. Once the quick scan is done, the majority of customers follow a very specific path. Most people will want to decide what looks good for an entrée first. So where do you imagine their eyes go first as a common place of focus? Did you say the right side of the page? If so, you are in the majority.

Studies reveal that there are two types of eye patterns when reviewing a menu. There are scan patterns and there are focus patterns. Basically we scan the menu to confirm predictability in the layout and find the category of interest that we want to focus on. We focus on those items of interest based on what we are in the mood for or what sounds good. That being said, the next question is, do people tend to focus in predictable patterns? The studies I have done show that the majority of people do go to predictable places on a menu and predictable places even within the area of a category. In Doerfler's study and in my study the same conclusion was reached. Below is an illustration of where the majority of people tend to focus when reading a predictable menu.

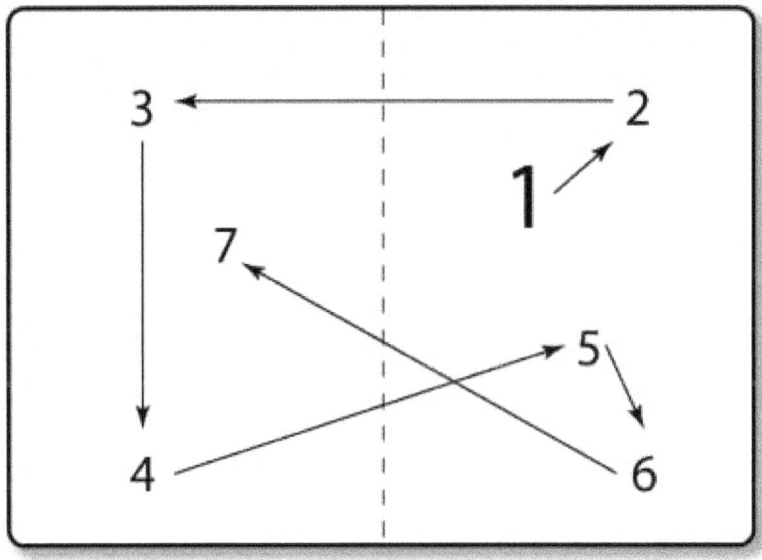

As you can see in the diagram from the studies this is the focus pattern of the majority of consumers once an initial scan has revealed to them that the menu is formatted predictably.

Here are two examples of how a single, double and a triple veiw menu are read.

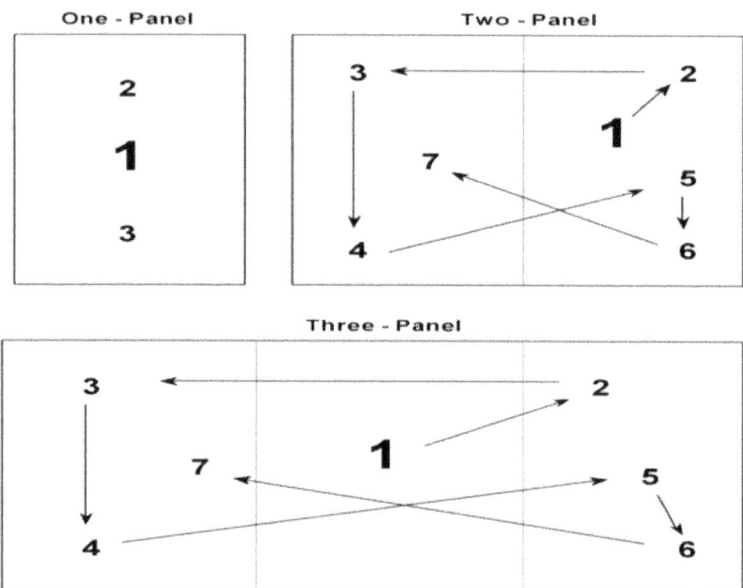

As you can see in the illustration above, the formatting has different rules based upon the view and the number of pages. So what does this mean to us when designing a menu? Well, quite simply, if we know where people are looking we can make sure we are showing them what we want them to see. In these categories I usually make sure to put the items that are most synonymous to our brand in the first and second traffic areas. Going back to the elevator speech I spoke about in branding (A.K.A. the two sentence speech) I will make sure I am putting the products that best embody our brand in these areas. As the sequence diminishes, I tend to put the categories in a form that follows the function statrting with 1 as the most important category and descending to 7 as the least important category.

In studying this effect it led me to wonder if I could take the process even deeper and explore a commonality in the eye pattern in a specific category and see within a category did people read top to bottom, left to right or did they have some other kind of process? What I found is that there does seem to be a common traffic path per category. I found most people's eyes read the first item in the category then they would jump to the middle and then the end. Once at the end, if they didn't find an item they were interested in or wanted to find more items of interest, the eyes would start reading from the bottom up. What's more I found that people often missed the second and the second to last item or read those items last.

That is not to say that every customer reads a menu this way. At the end of the day human nature is not predictable as a whole and there are always mitigating circumstances. What we are left with in these studies is to try and extract the most common patterns and do so from the most common audience. In studies done by Deloitte and the NRA every two years I have extracted that nationwide the average guest size or group size is 3.2 people per group. This research covers 8 years of data and thousands of restaurants, both full service and quick service, ranging from fast food to fine dining. With that in mind the majority of my research and conclusion comes from a test group of parties of 2-4 people dining out at both lunch and dinner. While certainly people in larger groups tend to follow different patterns, there is still similarity in focus per category but the scanning and focus changes based on the socail dynamic. Most larger groups (in a study of 6 to 20 adults, the highest traffic area and first traffic area is usually the top left (appetizers). The focus and scanning patterns are longer and tend to spend more time on items that are easily sharable. I will digress from this group size and the information as it pertains to the menu and follow up on some of these conclusions in a little bit. Right now I want to get back to the majority of findings and what we do with them in order to make menus more profitable and drive our guests to the items we want them to buy or at least give consideration to.

Knowing where people focus when looking at a predictable menu, we now have a tool for how to lay the menu out by category. So now by knowing that there is a category pattern I want to share with you how I attack that.

In a category I tend to focus on three items. The first three items I want to isolate are my three most unique items. I want unique items to be placed in the highest focus spots within a category for a few reasons. Again, those three spots would be top, middle and bottom. The reasons I focus on unique items are as follows:

- **To create a point of difference.**

- **To control the customer's perception of value.**

- **To capture more gross profit.**

An explanation is due to you on what I mean by all of these, so I will break them down for you.

Create a Point of Difference
I do not want my menu to be "me too" and I do not want my customers thinking I am just like the next guy. This one is pretty self explanatory but there are some caveats. I once saw a picture of a row of forks set behind a black background. In the middle of the row of forks, that were all aligned and the same, one fork was all bent up and misshapen and unusable. The caption below said:

UNIQUE – Just because you are unique doesn't mean you are useful.

I feel like this expression of imagery really hits home when it applies to a menu. I have found more success with aproaching a common, identifiable item that people understand in a unique way to be more impactful than displaying a unique item that no one understands just for the sake of being unique. Hopefully that makes sense. The NLP rule here is called pattern interupt and this rule is important for a number of reasons which we will discuss after this example.

EXAMPLE: A lot of menus have Buffalo Wings on the menu. This item is a really hard one to make unique. However, if this item was done in a different way, the idea of the buffalo wing is very familiar. An item I did on our menu at a highend restaurant was a korean BBQ, wasabi crusted wing. This item certainly had familiar ingredients both in the description and the presentation it delivered in a very unique way. We marinated the wings in this really dark Korean BBQ sauce then we baked them, glazed them and then tossed then in crushed up wasabi peas. When the item came out it was this beautiful fluorescent green from the crushed wasabi peas and dark from the Korean BBQ marinade. It also came with a side of BBQ sauce on a stringy nest of pickled Diakon radish. The item wasn't something people couldn't understand, it was just a combination they hadn't seen and thus we created our point of differentiation.

Here is another example. Many places have Poke or seared tuna on a menu. Just about everyone in the world (or at least my world) knows what a taco is.

So instead of putting a Tuna Poke or seared tuna on the menu like everyone else, we created a point of differentiation. Under appetizers, we put Tuna Poke Tacos on the menu. In the description I would explain that we make taco shells out of wontons and then describe how we season the poke. Again, this was an item that people understood but may not have ever seen.

The deep seeded reason for this diferentiation is more than just branding or making something remarkable. The menu technique we are going for is what is called "fair market value". Under the "Pattern Interupt" principle associate to "Fair Market Value" is a rule that basically says if you put a common item on a menu people will attach said item to a common price. If you were to ask the average person how much they would expect to pay for Buffalo Wings or a Hamburger at any given restaurant people would usually give you a price range. There may be variables based upon the concept type or even the size of the item but at the end of the day when all of those things are qualified, people have an expectation. To an operator that expectation is crippling. When you are relegated by the customer as to what is the fair market value of that item you are playing a dangerous game. If you are overpriced and not over delivering to the expectation you are not worth the percieved value the guest has of your product. One of two things happen in this situation, you either loose guests or you are forced to lower your pricing (or increase your cost on the plate to provide more value).

When you throw a wrench in the gears of "Fair Market Value" by using Pattern Interupt" through a point of differentiation, you have now opened yourself up to making more money.

How much should you pay for an appetizer portion on Wasabi Crusted Korean Style BBQ wings? Most people are not sure and surely cannot compare that to just any old chicken wing appetizer? And that my frineds is the answer you are looking for. In my Marketing Book "Whisper Marketing" we discuss in great detail the idea coined by Seth Godin on "Remarkable Makreting" the idea that anything worth remarking about is indeed remarkable. Most remarkable things are indeed fully vested in the idea of Pattern Interupt. So now you have two reason to do it… make it remarkable and make it profitable.

Let's move on to some more formatting rules and ideas for your menu. A basic rule of thumb I adopted from my good frined Tim Kirkland is in regards to soft drinks, iced tea and coffee. Tim Kirkland is an amazing public speaker and author of "The Renegade Server". In his book and passionately in his talks he details the idea of Iced Tea, Soda, Coffee and Water. He basically tells best thing that all of these things have in common. And that is…

NOTHING!

They all have a dimishing return. Hell, water even has a negative equity equation. After you factor the cost of the straw and the lemon you are loosing money everytime someone asks for water. Tim addresses this point in his book when talking about how to sell drinks and how to avoid selling these drinks. I apply that very same logic to the menu. Every customer in the world knows that you have water, iced tea, soda and probably coffee. I tend to leave these things out of the menu. The menu is valueable real estate and thus I do not want ot promote neglagible items or items with a deminishing return. People do not seek out these items too often so I don't think it will hurt you. Ask your server how many times people ask them in a day if they serve coke or pepsi products. What's more, listen when you are out or in your restaurant to the customer. How many times a day does a customer say they will take a Coke and the server will ask "Is Pepsi okay"? The same holds true for side items and salad dressings. People will often times ask what the chioces are instead of reading and certainly inspite of remember these items. If we know this rule to be genuinly true, then why would we muddy up our menus with this type on non helpful or otherwise unnoticed information? I challenge that you shouldn't. I also believe that leaving some diolouge for the service interction will also help more than hurt the dynamic of the guests experience.

Now lets get to one of the number one worst things you can do when it comes to menu formatting. Pricing Position.

Positioning in general has a lot of rules or guidelines to help you make the guest experience better and to make yourself more profit but this one above all else is the wrost thing that you can do and it looks a little something like this.

APPETIZERS

CHICKEN WINGS	$8.99
MOZZARELLA CHEESE STICKS	$7.99
NACHOS	$7.99
CHICKEN QUESADILLA	$7.50
JALAPENO POPPERS	$8.50

In the example above you can see the illustration of collumn pricing. The idea of column pricing is something we addressed just a ways back but to compound the issue you can see that when put pricing in a column you make it convenient for the guest to shop you based upon your price. Earlier I had mentioned that we need to format menus much a like a story. You may loose a lot of people when watching a movie if you were to give away the ending at the beginning. Column pricing does that exact thing. We need to follow an order just like a story line. We need a title, a plot and an ending. The title is your item, the plot is the description and the ending is the price. When putting pricing in a column you are promoting tha the reader jump ahead and spoil the ending. This formatt also allows the customer to skip the description altogether and gett right to the point. While you might think tha tis nice it certainly is no sales strategy. When you take an item you want to sell and then give it a unique twist, support it with a great description and then tuck the pricing in at the end you are creating a larger likelyhood that the read will start at the beginning and end at the end. If you do this the right way you stand a better chance to get the guest to buy based on somethgin more emotional or appealing and less logical or financially minded.

Other formatting rules as it appeals to pricing. In studies of scanning people are most distracted by the things that don't fit the norm. In studies of what captures peoples attentions in menus or what distracts people s attention in menu is aften times arrived at with the same answer. That answer is symbols. Symbols distract the read and draw there attention toward them. One of the most overused symbols in a menu is the dollar $ign. If you are in the slim majority of restaurants that wants to be known for it low prices or better than fair makret value than I would certainyl emplore you to make them a font size bigger and even bold or highlight them. If, however, you are the vast majority of restaurant I would suggest taking out dollar signs all together. Trust me, and you have to believe me when I say this, those little numbers at the end of your discrition withour a dollar sign will not stump your guest. They will know exactlly what to make of those numbers. Let's just not draw then attention to them any more than we have to.

In addition to the dollar signs make sure not to bold the price, make the price a larger font or make the font diffenrt in a way that will grab the customers attention.

In other menu examples I have studies I even see people put the pricing beside the item titem before the description. Again, follow the formula. Item, description, price.

PRICING RATIONALIZATION

There are quite a few moving parts as it relates to pricing rationalization. Before we jump into how to price and pricing logis, let's look at when to price.

I had mentioned earlier that the best rule of thumb for pricing is to follow the seasons. Four times a year is the best rule but two should be your absolute minimum. There is more to changing prices than just opening up your menu on the computer and adjusting pricing based on increased cost of goods. The number one rule relative to pricing changes goes back to NLP and fair market value.

Imagining you will or have heeded my advise to layout your menu in the order how we want a guest to read the menu: Item Name, Description then price you will still need to make a few more considerations.

Majority of the large menu thinktanks employed by the most successful reastaurant brands in the world, there is a common practice or two that must always be considered when changing prices.

As it aplies to NLP customer that have become regular at your restaurant have a recall center in their brain where they have modestly memorized your products, price and the location of items on the menu. Most big brands know this so when they do a menu change they move items within a category to different locations. They will also change the name or description of an item in a very sublte way so that the guest is clued into the change. The brilliance of changing the item's name or description slightly also affects the "Fair Market Value" portion of the guest prior experience. When they notice sublte changes they are typically more accepting of a price change than they would be if nothing changed but the price. Most brands will also typically promote a new menu coming in advance as well. Hopefully this all makes sense so far. Now lets look at some of the science and measureing tactics to consider when making pricing changes.

Reconfigure the sections.
You need to look at your items one category at a time and evaluate the winners and losers. Popularity is not the only consideration when evaluating an item. That item may serve another purpose like anchoring, brand image, remakrability, kitchen balance, and most importantly profitability. Here is a good illustration to use when looking at your menu items to gauge their impact to the bottom line.

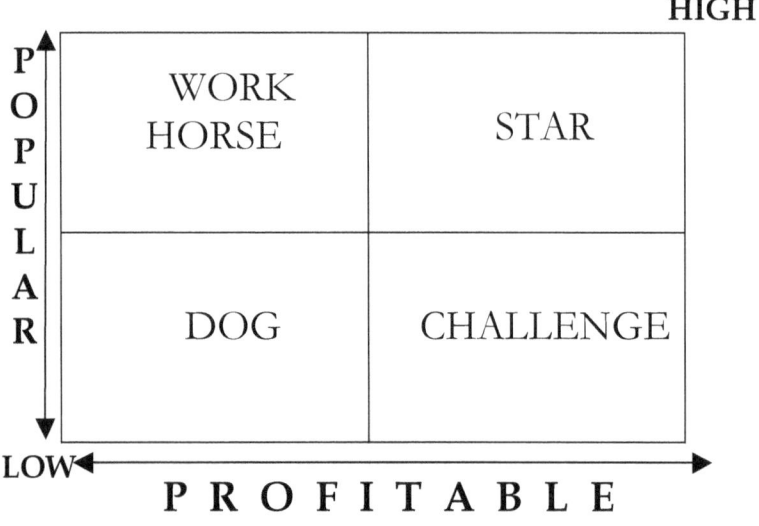

Looking at the chart you can see 4 quadrants based on a scale of high and low and popular and profitable. These are old advertising terms but they still relevant today. The table shows you how to measure an item and that cannot be done soley on the prfot or poularity of the item.

Take a look at the top right quadrant. That item is a star because it is both popular and profitable. Ideals we would want every item on the menu to be a star but that is not possible. Top left illustrates what we call a work horse. The item is popular but not profitable. Bottom right quadrant shows the challenge. This item is priftable but not popular. Last and liekly most important for the sake of menu changes and pricing exercises is the "DOG". Not popular and not profitable. So lets dig in to these 4 valuations of menu items so I can show yuou how to manage each on eof them.

STARS – These items are the best we can get. The big consideration when looking at these items is not to neccessarlity change anythgin about them but rather to find the similarities in these items and see if they can be applied to struggling items. Another consideration with Stars is to make them less visible. It is entirely possible these items owe a portion of their populatrity to the their placement. That being said, when you break down your items into these 4 categories you may consider putting a "Challenge" item in their place and moving the location of the star.

CHALLENGES – In reverse of the theory of Stars, Challenges should always be moved to higher visible places so that they be helped anlong to being a Star. Another ocnsideration when it comes to Challenges and a practice I will often employ is lowering the price. If the item is profitable but not popular, it may be perceived as not having a "Fair Market Value" and thus the reason it is not popular. The other reason I make price reductions is for those memorizers out there. I can say with a good conscious that "Yes, we did rasie some prices due to pricing changes in the makret and how inflation but the sameis true with items where we actually lowered some prices". Challenges are another item where I will focus on a little reengineering. I might play with the description to make it sound more appetizing and/or I might even make a sublte adjustment to the name of the item to make it sound more appeal, relevant and current.

WORK HORSES – This is an item I am going to do a great deal of engineering on. I will consider the name of the item, the components and the price and work to make it a star. I don't want to kill the item because it is popular but with it nor being profitable I may change the portion amount of products in the reciep to try and get it more profitable. If I do adjust down any of the portion amounts of items in a dish like this I will always rearrange the presentation to so that the value actually looks bigger. Here is a quick example. If I had a signature burger that was a Work Horse, I might move the burger portion size down from an 8 oz. patty to a 7 oz. patty and instead of putting the bun ontop and having the whole thing stacked I might leave it open face with the top bun off ot the side so that more of the plate is filled. Aslo with a new presentation the customer will usually be aware that something has changed. That said, I don't typically call out the changes I have made when reducing portion. Aslo, if I am goingto reduce a portion of something I will typically increase the portion of another part of the components. In the example above I might reduce the meat by one ounce but I will increase a less expensive ingredient like red onion mamrmalade from one ounce to once and half ounces. If you recall back in the "FOOD COST" chapter where I do the foodcost grid on an item I usually know the percentage of cost for the total plate that an item contributes and in my food costing program I can play with making adjustments like the burger example above so I can see how these changes will really impact the food cost.

NOTE: When making changes to a recipe you will always need to taste and make sure you are not compromising the overall experience of that menu item just to increase the items profit. Fact is, some items just can't be adjusted and they will always be lost leaders. If it is a signature item and you are not will to make any changes or take it off your menu then you will have to live it.
Remember, one item will not kill your entire

food cost. It is about the popularity, profiutability and overall blend of all items to measure your theoretical food cost. That is the number that matters!

DOGS – Dogs are usually the hardest thing on a menu to manage. I cannot tell you how many ridiculous arguments I have had with operators over these items. The numbers do not lie. I have litterally had operators refuse to take an item that they sell 2 units of each month off of the menu because they claim everyone will be upset. Here's the deal, if it doenst sell it is taking up precious real estate on your menu and potentially even watering down your brand. The reality is, no matter how emotionally vested you are in an item you can't force a consumer to like it or buy it. If you have dogs on the menu, here are your considerations. Rebrand the item! Start by trying a new name, change the portion or the qaulity of the ingredients, make it remarkable! Here is a quick example. Let's say your item is a Monte Cristo. I happen to love this sandwich but rarely is it done right. It also conjourse up an image of and oil drenched heart stopper sitting on the plate with a little rasperry jam. In todays environment it is not as magical as is was once in the eye of the consumer. But maybe you don't have to kill it alltogether. Image reengineering the item from a traditional greasy Monte Cristo (with a Diet Coke, of course) to something like this…

> *MONTE CUBANO ~ Pulled pork and shaved ham with Swiss and American cheese, mustard and pickles stuffed between cuban sweet bread, lightly egg battered and pan fried. Served with a side of honey habanero mustard...9*

It is still the essence of the idea but now it is reimagined and best of all, there is no real fair market value to an item like this because it isnot on every menu in America. So what is it worth? Exactally, you don't know and either does the customer but it sure sounds good and now instead of everyone ignoring you have given yourself a fighting chance.

Here are a few other considerations when it comes to Dogs. You can take them off the menu but that doesn't mean you have to stop making them. If the item is a dog but it is a dog of neccessity, meaning you have all the items on hand and that isnt going to change, then you still serve it. Instead of having it on the menu it is now a secret item! Trust me some of the biggest brands have made a mint in the viral word of mouth marketing of secret menu items. A few little brands like Starbucks and their Capt'n Crunch Latte or the Animal Style 4x4 at In & Out Burger, just to name a few. The idea is real easy, when a customer comes in to get that popular item you only sold two of per month, you tell them... "Yes we took it off the menu but for you, we will make it anytime you want it". Problem solved!

I hope I have illustrated the importance of taking a surgical approach to price rationalization when it comes to increasing your prices. Far to many times when people know that they are not making enough money they will just blindly raise all of their prices and often times by drastic amounts. This idea will perpetuate you problems by turning people off and thus pushing them away to come back with less frequency. They will aos begin to rationalize more negatively towards your brand. Desperate practices unmeasured and without justification will end you. I have seen it happen dozens of times. Please do not make sweeping pricing increases by erounious amounts and without making anyother changes.

Now I would like to show you some consumer study information when it comes to how a customer rationalizes your menu pricing. We have already talked about "Fair Market Value" and Formatting, tucking the pricing in at the end of the desctription but there are a few really cool tricks when it comes to the actual dollars and cents part of pricing formulation.

In a consumer prending study done a number of years back customers were given pricing on items and their reactions were measured. The best way I can explain this is in a reverse formulation I use for easy illiteration.

How much did you pay per gallon for gas the last time you filled up? Just about all of us will remember a ball park number, some will remember to exact penny. It is with those people I will ask them again how much they paid. Again, they will tell me and I will challenge that they were wrong.

The big reveal is that there is a third number right of the decimal point usually ending in a 9. If you paid $3.75 for gar you paid $3.76 because the gas was stated at $3.759. Yet we all miss that .009. It is with that in mind that we know customers rationalize pricing by rounding. There was a study done in consumer value decades ago that showed the importance of $.99 vs. $1.00. The value is true all the way along the scale. Just like the gas examples, consumers rationalized a value in things that ended in $.99. McDonalds is a great case study. Taco Bell, Wendy's and Burger King all have a $.99 value menu. McDonalds was forced into this gain and very much against their will. They wanted to stick to the profit formual that made them so successful which was the combo meal. Eventually they had to relent and they ended up adopting their own vaule menu. The "Dollar Value Menu". How much do you think that one penny meant to Mc Donald's? Likely a ton of money! This is because they say the value to the customer a little bit differently that the study showed. They could identify tha tto a customer having something on the menu for a "buck" was a real tangible thing in their mind. It might seem silly but the difference in a "buck" versus $1.35 wasn't something that was gogin to break the bank but it adds up to a $1+ more thant the other guys when you get 3 items. But McDonald's also understood to the customer there was no diffenrece in the value of $.99 vs. $1. It was a buck. People round up and down in their minds to rationalize price. Being armed with that knowledge a group of researchers from the Journal of Economic Psychology wanted to see what the threshold of value was and they hypothisized there were acceptable values and they varried within certain thresholds. Here is a summary of the data and what they came up with base on consumer feedback.

To a customer there was no percieved value difference in the following examples.

Item priced from $1 - $5. When these items were rounded up the nearnest quarter. So basically if an item on a menu was $2.13 the custoemr saw no change in the value if that same item was $2.25. The test was completed all the way through the range and everytime the item was rounded up to the nearest quarter there was no percieved change in the value of that item in the eyes of the customer.

The same test was tried from a range of $5-$10 and again they noticed some rounding strategies that didn't effect the customers percetion. They found they could round up to $.50 or $.99 on all items in this range and the value maintained its integrity. So if an item was $8.79 the value was equal when the item was $8.99 and so on.

The value was also done to a menu ending with a $x.95 strategy and it was changed to an $X.99 strategy. Again no change in value.

This may seem minor but if you have a menu where you are charging some random amount of change and not applying these rules you are leaving money on the table. The data tells us the customer doens't care so why should you. I have implemneted this logic in over 100 menus and I have never received an ounce of push back from the customer. It is worht mentioning that I have received plenty of push back from my clients but their customers didn't scoff. Imagine you could average $.10 - $.15 increase on every item across the board multiplied by the totla number of items you sell in a month or even in a year... how much would that be worth? You should still apply the previous rules to making such a sweaping change but this one really blew me away when I first read the study and started to apply it.

So what about items that are priced above $10?

Good question! This study changed quite a bit in this part of the value perception. The data was not as conclusive but the majority of people could rationalize a price increase all the way up $XX.99. However when a $10.99 item went to $11.00 there was a value change with most customers. The conclusion of the rule was to round any double digit item up to $.99 and the value was still there. However that one penny that would make the item change from $10 to $11 was very notciable.

The study continued to test the value if they didn't put any pennies on the end of the price. The data was surprising at first but as I explain it to you today it will make absolute sense. When an item was priced with no decimal point of pennies it was percieved as a higher quality item.

The difference would look like this…$10.99

Versus the change to this…$11

It is common place today and that is the strategy that most fine dining concepts use on their menu. It is not for me to say which came first… the chicken of the egg but I eat eggs for breakfast and chicken for lunch so to me, it was the egg and I would apply the same logic to this menu idea. Fine dining restaurants have been using this practice for years and I believe that the customer associates a higher value to a pennyless menu item because of the restaurant business as opposed to the restaurant business doing it because they knew of this perception.

It was with these studies and this applied science that I came up with my own study and application. It has been years since I have put a dollar sign on a menu because I know that symbols attract peoples attention. We have already covered that but I wanted to play with some pricing symbols because use to play semi-professional poker. It was never my full time job but I was sponsored and I made enough money to quit my job and be whole, problem is I have always loved this business more than anything else so I chased this dream and enjoyed that hobby! Sorry for the sidebar but there is relevence to this story. When I first started playing poker I remember a table discussion one time discussing the use of poker chips and the associated value to money. The casinos and gaming in genral realized early that ot a player chips were chip not money. There was a disconected value in the currentcy when it was changed. It was worth the exact same amount but it rationalized comepletely differently. It was with that logic I decided to play around with fractions. Here is what I found out.

In a samples of 3 different case studies in 3 different states along thee different concepts types the results were all similarly alligned. My test studies were a fine dining concept, an Irish Pub and a gourmet taco concept. All three of the concepts had one thing in common. I would define all of them as edgy or at a minimum on the cutting edge in their enviroments. Basicallt these were not some cookie cutter generic concepts. The consuer value in these case studies seemed to have a disassciated value from the money. People didn't see the vlaue as high or low but rather as something remarkable and worht talking about. When asked about the value most of them seemed indiffernet eventhough in all ofthese case studies in their respective market these places were all priced above the average concept type that was siminlar in their market place. People didn't focus on the price but rather the cool factor. It was a remarkable thing to them and worth telling people about. It drew a lot of conversation at the table and even afterwards. The items on the menu would have looked something like this.

MONTE CUBANO ~ *Pulled pork and shaved ham with Swiss and American cheese, mustard and pickles stuffed between cuban sweet bread, lightly egg battered and pan fried. Served with a side of honey habanero mustard…9 ¾*

I only use pracftion to the nearest quarter as not to make peopl ethink tho much but the experment ended up leading me to capture more money that I would have if I had uses pennies. I could actually get up to $.75 more out of an item than I otherwise would have following some of the previous rules. Now, while I know this logic is not for everyone and to date, even with the imperical, statistical data I have, the social proof and the testmonials from the owners of these concepts, peopl eare still reluctant. I am not forcing this one down anyone's throat but if you think you are cool enough to employ this strategy... DO IT! When you get the same results I did, send me an email and tell me all about it! info@profitproplus.com

I am currently working on some other pricing studies and I wll make sure to post them as soon as we have some definative data to share.

In two markets right now I am playing with an Irish and an English pub putting the pricing in Euro's. It is a little dangerous and needs to be called out but it fits the concepts and I believe we will be able ot breakdown a lot of "Fair Market Value" thinking with this one.

I am also playing with categories based on price. The entire category is one price and the price changes as the quality of the category goes up.

We are also doing some Market Value pricing studies in places on the menu where they absoelutely do not belong just to measure the reation and see if we can find some wins. Stay tuned!

THE RULES

You have to learn the rules of the game. And then you have to play better than anyone else.
Albert Einstein

I am not a huge fan of rules, as I mentioned earlier but that is not to say pure chaos is okay. I am a firm believer in guidelines but as is the case with all rules, even the ones I am giving you, they should be challenged, improved upon and reinvented. I try to keep up on these rules but almost inevitability every time I publish something I think about something I missed, should have gone into more detail about or learn something new that I wish I had known before. That being said you should use this chapter as a quick reference to some great do's and don'ts that will make your life a little bit easier and at a minimum provoke some thought. So without further ado…

KIDS

Here's a random factoid to consider…

Children drive over 60% of the average family's dining decision. This is not so say that kids are making the decision but rather that the decision is made with the kids in mind.

What are you doing for the kids?

If you think chicken nuggets, a grilled cheese sandwich and macaroni and cheese is the right answer you are missing a huge opportunity.

Your menu should be geared specifically to kids. They should have their own menu. You do not have to have coloring crayons to have a kids menu.

If you are looking for affordable full color kids menus that look awesome and won't break the bank check out kidstar.com
They also have a lot of other great kid's product at tremendous prices.

Check out the idea section for some cool ideas I have done or seen for kid's menus. The little factoid isn't meant to impress upon you how happy the kids are, while it is important that they like your food you need to focus on environment as well. Do you have games, books, TV's, a separate kid's area, etc.? When kids are distracted, or occupied, as the good parents like to call it… parents are happy.

Happy kids = a higher likelihood of getting the family back more often.

Kid's food specifically catered to kids is something I did at one of my family restaurants. We served spaghetti and meatballs in a large soda fountain style parfait glass with parmesan cheese and a cherry tomato on top. We had a picture of it in the menu and the kids loved it. We also did meatloaf cupcakes that had mashed potatoes piped on top of them with a cherry tomato on top. We also had a build your own pizza and build your own burger section for kids. They were very engaged and much like adults when you have a unique experience you tend to remember it and this is exactly where we want to be when people are decided where to eat... top of mind!

PICTURES

Here's a rule that will likely make all of the sense in the world. Using pictures of food on a menu is a very dangerous proposition. As a rule of thumb this creates a number of issues and because of that I will avoid using pictures at all costs. The few exceptions to the rule usually only apply to quick service concepts.

Here is the challenge that pictures create.

Quality – Pictures are proven to cheapen the experience which diminishes your fair market value. You likely do not think of the highest quality restaurant when you think of a menu with pictures. Poor quality images will also send a bad message. If you are going to use pictures they must be professionally taken with proper lighting and turned into high resolution images with quality printing. This is usually cost prohibitive and forces and entire menu change when reformatting.

Real Estate – Pictures can falsely enhance the perceived size of a menu forcing you to have to have more pages with less content. This is exhausting. You do not want to wear a customer out when they are trying to make a choice for dinner.

The Right Picture – While it is proven that pictures to serve a purpose: pictures on a menu usually equate to those items outselling other items. This is meant to be very intentional so if you must, make sure it is the right item.

Over Selling – This is the most common culprit of pictures. Typically when an item is made for a photo shoot it is meticulously crafted, embellished and well lit. Don't believe me… just go to any fast food place in the world and look at the picture of the product they are selling you. Then order it and hold it up to the picture. If you can visualize this or this is a phenomena that has never happened to you, you will see what I mean when you do this little exercise.

FONTS

Limit the number of fonts you use in a menu to 3 or 4.

Use fonts of at least 12 points in size or larger.

Use fonts that are legible. A fancy font might look cool but if it is hard to read you will confuse your guest.

ROMANCING THE MENU

This is a term we use to create a point of differentiation. Using adjectives is a great way to help paint a mouthwatering picture in the mind of you customer. The only caveat in this idea is getting to wordy or worse over selling an item that cannot be delivered on. While it is great to embellish a menu that delivers there is no greater risk of disappointment than making something sound better than it actually is.

One way I have found very successful in romancing the menu it to actually take common items and upgrade them. When I spoke earlier about fair market value I told you to stay away from "me too" items on your menu. To take that idea in a different direction here is an example of how you can have a common item but change the perception of fair market value.

Look at the example of this burger…

LAME
The Me Too Burger – topped with bacon, Swiss, lettuce and tomato…8.99

ROMANCED!
Gruyere Maple Bacon Burger – Our signature custom grind burger with melted Gruyere cheese, maple bacon, arugula and vine ripe tomato with sriracha mayo…9.99

In the romanced version above you can look at the perception value and likely see that the example is easily worth the $9.99 whereas the "Lame" version likely seemed overvalued. I was able to get a dollar more, have a better perception and the total cost to plate would be about $.20 more cents. I simples brushed my same bacon with maple syrup before I baked it, I upgraded from Swiss @ $.12 per slice to Gruyere for $.15 per slice and I added arugula with is about 20% higher than lettuce. The rest is all just a more eloquent description.

Ingredients can change an entire perception of the menu in the eyes of your guest and it is really as simple as the example above. Language is also critical in romancing the menu. Some words have a negative connotation or provoke imagery you may not want on your menu. Here are a few words to stay away from. Let these words bounce around in your head for just a second and do a little word association with them as you read them. What is the next word that comes to mind when you see this words.

Moist.

Supple.

A pile of…

I don't want you to get you too paranoid about all of your language but be mindful of what words provoke what kind of association. Don't use language like deep fried when lightly fried or flash fried would get the same information to the consumer without conjuring up imagery of gross, greasy food. Other words to be cautious of are words that sound heavy, make people think your food is unhealthy or bad for them…

Smothered in…

Drenched in…

Buried in…

Hopefully these little examples give you an idea of what I am talking about.

Look at your competitions menus and see what it is you like about their descriptions, their language and their delivery. Does it have a good value? Did it paint a good picture? Did it over deliver?

The last thing I will say about language is the subtle art of the under delivered description. Higher end concepts have been doing this for years. They will simply list the main ingredients like a list and when the product comes out it goes way beyond over delivering. In a high-end or white tablecloth concept this can work really well. It is received as artsy, understated and edgy. Make sure you know how you are seen before executing a task like this. If you are going to understate the item you really need to over deliver it and make sure you have a few secret ingredients that make the dish pop that come as a pleasant surprise.

Here are a few restaurants that I think do it really well. Check them out on line and you will see the value and point of differentiation these menus create.

Marcos Coal Fired Pizza, Denver

Gabbi's Mexican Kitchen, Orange

TAG Restaurant, Denver

La Puerta, San Diego

Urban Plates, Irvine

The Meatball Shop, NYC

Animal, Los Angeles

Rubicon Deli, San Diego

Mendocino Farms, Los Angeles

These are all great example of menus where you will see a big difference in romancing the menu.

ANCHORING

This is an NLP term but it works magically in the world of restaurant menus. Anchoring is an idea that leads someone to think something positive, even if the anchoring device is negative.

In a study done on menu pricing, Anchoring is a

technique used to show value. The study basically illustrated how a high priced item was seen as over values. I believe the example was illustrated by an $18 frittata on a brunch menu. The item was a really good item but it didn't sell. It was basically more expensive than any other item on the menu. Here is an example of what that item looked like.

> **FILLET MIGNON FRITTATA** – *Egg, Fillet Mignon, Heirloom Spinach, Brie Cheese, Tricolor Peppers…18*

The example as it sat in the menu category was over $5 more than the next item on the brunch menu. Because of that people thought the menu was overpriced because they didn't see the value in the item as it related to the other items. Even though when you look at this item as described above, when it is not in context and you are not looking at the total menu you might see some value in that item. Then again, if you are like most people, paying $18 for what is basically a steak an egg omelet doesn't seem to be reasonable for what we basically think of as a breakfast item. So here is how they used Anchoring to change the perception. In the middle of the "Brunch Classics" section of the menu they put in their anchor. The anchor is an item intentionally meant to grab people's attention. In most cases people will use an anchor to create a huge value by making that item stand out and be such a low price point that compared to the value people might think the restaurant is crazy or basically giving the item away. In this example, the operator understood the value of anchoring but used it in a different way. The item that was anchored was as follows...

> **SURF & TURF FRITTATA** – *Kobe beef tenderloin, sautéed lobster tail, 3 year aged parmesan, sautéed white asparagus and topped with beluga caviar...$50*

The anchoring idea here was used in reverse. They wanted to show how much value was in the first example of the Fillet Mignon Frittata by showing people a real crazy example of the "Expensive Item". And when people saw that item on the menu, while it was hardly ever ordered the value of the Fillet Mignon Frittata was received as better value and it proved in their sales. The Fillet Mignon Frittata witnessed a huge spike in sales and in that category went form the number 3 worst selling item to the number one bestselling item. The Surf & Turf was the worst selling item in the category and that was the point. Often times the restaurant was out of stock on the item because they were out of the beluga caviar.

I know how crazy that sounds so I will give you a more practical idea that is often used for anchoring.

This example lives in the wine list. People find a value in wine based on the pricing of the wine list. If you most expensive wine on your list is $55 people will often order a wine that is second or third in value like a $50 or $45 bottle. In multiple studies I have conducted I found that by simply putting a few more expensive wines on the wine list the average bottle of wine sold went up by over $12. Hopefully you have been there. Maybe you have been out on a date or at a big meeting and you decided to order wine. You don't want to look cheap so decide to find a medium or medium high priced wine so that it looks like you are not being cheap but also not showing off. If that resonates with you, then you have been a victim of anchoring. Take a look at your menu and see if you can implement some anchoring strategies in your menu to create more value. It is a beautiful thing of social science!

HALF VALUE

Anchoring is a perfect lead in to the next rule. The rule of half value is also relative to perceived value and one that blew my mind the first time I saw the study.

Earlier I explained how the guest sees more value in the depth of categories than they do in the total number of items.

Basically you can have fewer items on the menu with more categories and the guest interprets a better value. A good thing to know when building a menu and the rule of "Half Value" can be a critical tool for expanding you categories and knowing where to do so.

The rule of half value basically says this. If one item is more than double the cost of any item in a category it will lead the consumer to think you are overpriced.

Here is how it works. Let's take a look at a category like appetizers. Let's see what you think. Really study this one and see what your impression of the value is.

CHIPS & SALSA
A pile of tortilla chips and salsa............................$3.50

MOZZARELLA STICKS
Six sticks, breaded and fried, with marinara sauce........$6.50

SUPER NACHOS
Heaped with ground beef, cheese, guacamole, scallions and black olives.........…................................…......$11.00

POTATO SKINS
Four skins stuffed with potato and cheese and topped with sour cream and chives…....................................…………..$7.00

GRAND SLAM WINGS
Can you handle it? Hot, Extra Hot or Fireball. With ranch dressing and celery sticks..............................…......$10.50

EXTRA CHEESE FRIES
A huge plate of fries smothered in cheddar cheese…....$5.50

CHILI CHEESE FRIES
Chili and cheddar cheese over steak fries. Ask for sour cream on the side...…..$6.75

In the example above most people thought the super nachos seemed overpriced. Most people didn't even read the description. They skimmed the items and the pricing and they thought the value is misaligned. The first mistake was obviously an illustration of why you should not do column pricing. As we discussed earlier, when you list your pricing with dollar signs and put them in column pricings it allows people to find value in the menu based on price and "Fair Market Value". So what does this have to do with "Half Value"? Well it is quite simple, much like anchoring but in an opposite way, having an item that is way too low in the over value scheme of a category can make the other items look over inflated... even if they are not.

In the example above, if I were to pull out the $3.50 chips and salsa, the items in the category would look to have much more of a value. While there are a lot of things wrong with this example, the "Half Value" rule makes the biggest impact. The nachos could have been romanced and certainly could use an adjective relative to size to create more value. The pricing could have been tucked in and void of the $. The pricing shouldn't be bolded but the biggest issue is having such a cheap item in a category that was less than half of the next item.

Now, going back to the rule of categories and adding more categories to create more value, I will use this example above as the exact reason why I would split these categories.

I would create two categories from Appetizers and change them into something like this...

LITE BITES

CHIPS & SALSA
A pile of tortilla chips and salsa...3.50

EXTRA CHEESE FRIES
A huge plate of fries smothered in cheddar cheese...5.50

CHILI CHEESE FRIES
Chili and cheddar cheese over steak fries. Ask for sour cream on the side...6.75

MOZZARELLA STICKS
Six sticks, breaded and fried, with marinara sauce...6.50

FOR SHARING

SUPER NACHOS
Heaped with ground beef, cheese, guacamole, scallions and black olives...11

LOADED POTATO SKINS
Eight skins stuffed with potato and cheese and topped with sour cream and chives...10

GRAND SLAM WINGS
Can you handle it? Hot, Extra Hot or Fireball. With ranch dressing and celery sticks...10.50

I didn't really change much of anything. I did increase the potato skins portion amount and the price to balance the two categories other than that, now the categories create more value. Adding the category "For Sharing" the size of the portions is implied even though all I changed was the potato skins. The size or value of the lite bite is also implied because as words "Lite" & "Bite" usually denote an association equal to something less expensive, healthy quick, small or an amount to just tide you over. This is an actual before menu from a client of mine and like I said earlier, I would change a lot of things in this menu (and did) but the illustration of "Half Value" is really all we were going for here.

SATURATION POINT

The next natural progression of rules to talk about is the rule of "Saturation Point". This rule ties perfectly in to "Half Value" and "Category Rule".

"Saturation Point" is a rule that came from a study the size of categories and how engaged people are with a certain category. When a category has more than 9 items in the category it is typically overwhelming. This is true for a number of reasons. The two most important reasons are this.

Remember, we scan top bottom and middle, then

we go out form the middle with usually the last placed being second and second to last. In a category of 9 items the last two items to been seen are the 3rd item and the 7th item.

The second reason why 9 items or more starts to become too many is because of specificity. Typically when there are over 9 items in a category that category can usually be broken up in to more specific categories. This saves the reader time when looking at specific categories as they skim. They can identify the category of interest and focus on that category faster, more efficiently and effectively.

NOTE: Remember, the average person only spends about one and a half minutes reviewing a menu. They get frustrated when they have to spend more time than that and they will usually end up just playing it safe out of frustration. Playing it safe will not allow you to stand out and will make the customers experience less than impactful.

So what is the ideal number of items in a category? The ideal number of items in a category is 5 to 7. Having an add number of items makes scanning easier because there is a more defined middle but the science of that part of the study is not definitive.

COLOR RULES

We discussed color and color psychology earlier in

formatting but here is the rundown on some simple color rules to remember.

The BLUE rule!

Blue is the only color that does not naturally occur in food. Blue is most closely tied to a prerecorded rule handed down to us a survival instinct. Given that blue usually occurs in food with mold we are hard wired to receive the color as unappetizing. When it comes to food advertising and menus it is a good color to stay away from.

The EARTH TONE rule

This rule is really simple, we receive earth tones as being natural and that is a great stigma to have on a menu. Green says fresh, brown invokes images from the land like cattle. Most earth tones are colors we see in nature and there is nothing better than be perceived and natural and wholesome. Burgundy is another great color. It is an earth tone associated to wine and wine is associated to wealth. Thus burgundy is associated as a "rich or wealthy" color.

The RED rule

Red stimulates and excites. Red is an action color but it can also be a threat color. Red is the color of blood and invokes emotion of anger as much as it does create excitement. It has multiple associations. Red is also a color of "fast". Red cars get more tickets than any other color of car (I've heard, can't substantiate because the research is not that important to me).

The YELLOW rule
Yellow is the color of sunshine and "Happy". This color stimulates and excites. Both which are great imagery associations. Yellow can be hard to read and also seem desperate or too bright like you are yelling to try and get attention. When yellow words are on a black background they pop really well. Yellow and black are the highest impact marketing colors used in advertising campaigns. Think McDonalds! They know exactly what they are doing with color.

The ORANGE rule
Orange is soothing color and the most universal gender neutral color. Most colors are gender dominant. Blue for boys! Pink for girls! Orange is one that has no associated value thus it makes for the best neutral gender marketing color. It is an action color that stimulated appetite and appeals to both genders equally when trying to stimulate a sale. The earth tone versions of orange have a combination effect.
   (See Earth Tone rule above)

The PURPLE rule
Purple is very similar in rule to burgundy. It is rich and elegant in interpretation.

These are just some of the more common color rules. Think about how you apply these to your menu to create a positive reaction in line with your demographic. Try to minimize using more than 3 colors. It is also wise to invest in a color wheel or look up how to use a color wheel on line. Color combinations when done properly can create a nice calm. When done improperly they can create confusion.

Remember, in all of the rules the number one impact we are trying to avoid is confusion. "A confused mind does nothing".

HIGHLIGHTING RULES

Highlighting is a common menu practice that has been used for years. Now that you understand many of the attraction rules you can use negative attraction in a positive way. The age old way of attracting attention it through "boxing". That is to say, creating a box around an item that you want to draw attention to. The tradition line box usually looks very desperate and typically doesn't follow a form in line with the design. You can create highlighting by putting an entire color around items you are trying to highlight. Here are some other ways to highlight items.

- ✓ Use a bigger font.

- ✓ Use a different font or capitalization scheme with the letters.

- ✓ Use a different color. (Go back to the color rules)

- ✓ Use format change to highlight an item. Indent or change form.

USING SYMBOLS

This one requires some outline. You now know that symbols grab people's attentions. You have probably fallen victim of this in the past when have been out to eat. You need to use symbols in a way that makes your audience wonder what the symbol is.

Have a key that is easy to find and use it intelligently. Don't use too many symbols or one symbol too often or you will water down the impact. "Our Favorite" as an answer in a symbol key is worthless. Most guests will think you are sending them there because it is a more expensive item or more profitable and with this it creates skepticism. Try using Symbols to promote an award winning item or a new item.

NOTE: Award winning is very ambiguous. You can have a competition in the kitchen to see who can come up with the best new item and the winner gets an award. See where I am going?

Using a symbol to denote spicy or gluten free is silly unless you are in that kind of market specifically. And if you are in the specific market niche, the symbol should go without saying. Put gluten free or spicy right in the menu description. Use symbols to truly impact the masses with an item that is a true point of differentiation.

MENU COVERS

We also talked about this rule a little in formatting but I want to impress this just a little bit more.

Get away from the traditional menu. If your menu design cover looks like every other menu you will be relegated to be seen by the guest as nothing special or just plain ordinary... just like every other restaurant. You do not want that.

As a best practice...

Do not laminate your menu. Get creative and find a better way.

Form need to follow function. Make sure your menu is easy to tweak, reprint, adjust and manipulate.

Keep it clean. Literally make sure you have a menu that you can clean or cheaply dispose of when they get greasy or dirty. The last thing a guest wants is dried salsa from the last guest on their menu.

Make it unique but affordable to change.

EXAMPLE: Bouchon – Las Vegas: A restaurant by Thomas Keller uses a newspaper thin, craft (brown) colored paper to print there menu. It is folded up around their napkin and silverware. It is unique, affordable and easy to replace. Ideas like this really stand out.

MENU DEVELOPMENT

"In order to make an omelet you're going to have to crack some eggs."
- Unknown

We make menu changes in our restaurants twice per year. Here is the formula we use when identifying what to change and how.

The first thing to call out is that we have a process that is documented and scalable. Here is our process.

NEW MENU DEVELOPMENT PROCESS

- Menu Evaluation/Change Development Meetings – March 1st and August 1st
- Menu R&D – March 7th & August 7th.
- Menu Tasting and Evaluation – March 14th and August 14th.
- Menu Final Approval and Roll Out Process March 21st and August 21st
- Pre-launch Marketing, Branding Messaging & Training March 28th & August 28th
- New Menu Launch – April 1st & September 1st
- Measuring and Recording - Ongoing

Here are the rules and considerations we use to evaluate our menu changes and initiatives.

THE QUESTIONS and RULES?
WHEN?
Q: When do make these changes?
A: First we begin with our Menu Evaluation/Change Process. We start working on this process one month before the new menu implementation. We specifically set our new menu dates for September 1st and April 1st because we find that our guests view these dates as the transition to Fall/Winter and the transition to Spring/Summer. We also know that this is when the seasonal availability and pricing of food begins to change. Because of that we start our development process of the new menu 1 month early.

WHY?
Q: Why are we looking to make menu changes?

A: For two reasons. We want to keep our food fresh and we want to keep our ideas fresh. For example, we know that the human mind has two big shifts in eating behavior through the course of the year. When entering the Spring and summer months we look to lean up and get our bodies in shape. We want to look good at the pool or beach. Plus the psychology of the mind tells us when it gets warmer we do not need to put on the necessary fat reserves to get us through the cold season. The inverse is true when the cold season comes around. This goes back to the our evolution but is something still hardwired in our brains. Beyond the behavior is also the consideration of food availability. As the seasons change so to does the availability of different types of fruits and vegetables. With this change comes a shift in the supply and demand equation of commodity items and that impacts the pricing. We want to take advantage of these changes on our menu to eliminate the items that are going to go out of season and thus take large price increases due to shortages in availability (and quality) and shift to the items that will be more abundant thus driving the prices down.

A: We also want to stay relevant. We want to keep our ideas and offering fresh to our guests so that they do not get board with the same ol' same ol' while still preserving the things they have come to know and love.

As a subset of keeping our ideas fresh we have to look at a couple of things.

We want to know what is trending. It could be a new food craze we might want to consider.

We need to consider the success of a new competitor that has entered into our market with popularity and identify what items might be doing well for them that we had not considered. As long as these items are true to our brand and our concept we will consider looking at how we can menu them to fit what we do.

We need to look at health trends and crazes that may be impacting the market place.

We will also need to review any new food rules, laws or popular cultural impacts as it relates to food.

Next we need to know the who, while still considering the answers the following questions

WHO?

Q: Who is our audience or demographic? Is the disconnect in our low performing items due to our audience? Are we trying to be what we want to be vs. what our clientele has told us they want?

A: We are very careful to listen to what our guest tell us, even if they do not tell it to us directly. We look at the data to get the answers and we look at what our guest like, think and do to help us serve their needs. When have found by taking this kind of an analytical approach to our business and by using this type of awareness not only do we react well but we also anticipate very well. We do have one caveat, we have to be true to our brand first, we know we can not be everything to everyone and when restaurant try this philosophy they often times create too much confusion and water down their brand, their value and the trust of the guest.

Next we need to know the what.

WHAT?

Q: What do we know?

Q: What could we improve or what isn't selling?

Q: What do we do well or what is most popular or profitable?

A: We begin by looking at our historical sales and through our food costing and menu engineering program for data and reporting. We use Profit Pro Plus software and the reporting tells us what items are most popular and what items are the least popular. First we assess what are the least popular items and we put those items on our list. Next we look at what is the most popular and put them in a separate column on the same list or on the white board.

Next we review our list with the team to try and figure out why our least popular items are that way? We ask our selves a series of questions like this?

- o Does this fit and enhance our brand image and what makes us better, different or special?
- o Is it profitable? If it is we keeping digging, if it is not, we consider replacing it.
- o Is it price? Does someone else out there do it for less than we do?
- o Is it quality? Does someone else out there do it better than we do?
- o Is it value? Does someone else out there do better and cheaper than we do?

- o Can we do it better by rewording the offering, placing it in a different location?
- o Can we reimagine the item to make it sexier or more appealing?
- o Can we reengineer the item to match up closer to the items that are popular?

Last we look at the how.

HOW?

Q: How do we roll this out?

Q: How do we avoid pushback?

Q: How do train to the new items?

Q: How do we respond to the guest?

A: The explanation of how we roll this out is listed above in the overall outline. **As an answer to the rest of our questions,** we avoid push back by promoting the menu changes well in advance. We put up flyers in the store to build the new anticipation. We run marketing campaigns around the new menu launch with a call to action and prizes when people come in a try the new items. We build language to explain why we made a menu change. We will explain to our guests that the items became cost prohibitive due to availability and we feel like we could pass along the price to the guest because it wasn't in our pricing structure to provide great quality product at a great value. We may also site quality issues we were having with the product, etc. Either way, we anticipate the guest negative reaction and we build and train our teams on delivering the message. We also will make sure to have some product in reserve for a short time or consider keeping the item but no linger showing it on the menu. At which point we will explain to the guest…"I'll tell you what, it is not on the menu but we have a new item very similar to what you might have loved about that item. So if you would consider trying this new item, and if for any reason you are not happy with it, we'll buy it for you and let you try something else. At the end of the day we understand we have got to break a few eggs to make an omelet and we are not going to keep every guest happy but if I am eliminating an item for any number of reasons and I sold so few of them that it made sense to get rid of it, I have to be prepared for some pushback. However, we can't be everything to everyone so I have to take some risks. And that leads to a much bigger question… Do we feel confident that these changes will help us attract more than we stand to gain?

Once we have identified the items we want to impact, change, modify, add or eliminate we need to get to work on the culinary side. Our culinary team has their own set of rules and questions they have to consider before getting started. Here is what we look at from the culinary side if the isle. Here is an actual example.

Q: What are the rules of our brand or what do we stand for?

A:
- o We are a lifestyle brand for active, creative and casual people.
- o We offer hearty food with bold adventurous flavors.
- o Our food is always fresh and made to order .

When making our new menu items they have to first past these rules above. So we ask ourselves…

- Does this item appeal to the active, creative and causal audience?
- Is it a hearty item with bold adventurous flavors?
- Can it be made fresh to order?

Once we can check off this list we have other considerations to look at.

We consider a decision Delta.

Do we **Add** - **Enhance** or - **Eliminate** the item? We make the decision based on the current performance of the item, its profitability and its uniqueness to our brand. There are always some lost leader items that have to be considered.
(look at our other decision deltas at the end of this chaprter)

Next we look at some of the financial implications vs. some of the benefits of the item.

Does a new item introduce new inventory we can use somewhere else? If so great. We try to use a rule of 3 when bringing on an inventory item. If we can use it in the development or modification in multiple items that is great, if not we try to think around the idea or limit the inventory to dry ingredients only so we will have shelf-life and practicality to adding the item.

We also want to ensure the items have availability throughout the coarse of the menu items life. If it is a seasonal item, will it take us up until the end or our respective season: September 1st or April 1st? Many times we have found gaps in this questions and that has forced us to rethink the product either do to lack of availability, diminished quality or an end of season price spike.

When we do like an ingredient that we can not use we will consider a different rule according to Neurogastronomy. What the?

Yeah, Neurogastronomy, here's how it works. We understand that a lot of food at it's very core value has rules from nature that it follows as well. In the science of food we look to the PH scale. We know that items are classified on a scale of 0-14 and have 3 different degrees of classification. We have an acid, a neutral and a base (or alkaline).

If we find an item that we like but is not feasible to use we apply these rules to the item.

Is there an item similar to the one we want that is more feasible that matches the desired outcome?

- **Is the texture there?**
- **Is the consistency there?**
- **Is the PH balance of the item in line with the item we like?**
- **Is it available?**
- **Is it affordable throughout the term of the menu items life?**

If it passes this test, we will try it our and see if it does the same thing for our recipe as the other item we wanted. Sometimes it doesn't work while other times it surprises the heck out of us.

Once we have our menu ideas locked in and we are ready to start testing them we move on to 5,4,3,2,1 methodology of Crave-ability for a new menu item.

CRAVE-ABILITY RECIPE RULE
5 Pallet Flavors Zones in the Mouth
  (sweet, sour, salty, bitter/heat and umami)
4 Need to be satisfied
3 C's (Creamy, Crunchy and Chewy)
2 Temperature Profile (Hot and a cold)
1 Successful Recipe

This is our science but one that we have studied in depth and has proven very successful for us. When a recipe item follows tis formula we have had greater success with it than any other grouping of factors. We have layers of flavor, layers of texture, layers of temperature and that creates **crave-ability**.

After the items are successfully put through R&D we are ready to taste, discuss and vote.

When an item makes it through committee we now have to work on the messaging and brand elements of the menu.

First we need to name it. We have naming committee rules as well. Here is what we look for?
- **Is it unique?**
- **Is it familiar?**
- **Is it desirable?**
- **Does it make sense?**
- **Is it creative?**

These questions are going to be specific to your menu, your proposition, your brand and your audience. Here is an example of an item we did.

In our chicken wing concept we new we needed to take some relief off of our tradition and boneless wings. As the market conditions continue to change against our favor we need to stay competitive but knowing we are a wing concept we also needed to add a diversion within our same category.

Our wing prices had led us to upwards of a 35% - 38% cost on our items. We could withstand a small price increase but not enough of one to get our profitability back in line with our standards. We obviously couldn't eliminate the items as they were our namesake so we decided to alleviate some of the pressure. We added a few more categories to the menu to relieve the pressure.

We added a sandwich category that had very little impact on our inventory and only had to bring in the 3 dry ingredients which we were able to use throughout the menu in at least 3 applications. But it is what we did with the wings that really shifted the impact on the menu.

We served traditional bone in wings and boneless wings. We sold our wings in increments of 5 for bone-in and 10 for boneless. Either style and respective count sold for $5.00. This was aggressive but that is also what got us in the game.

After going through our R&D process and answering all of the questions and using all of our considerations we decided to add a new category of wings. We call them crunch wings.

We made 3 new signature recipes in this category. These wings came with a sauce you could only get on this wing, not available for any other application. They also were prepared with a crunch crust on the outside. Here are the items.

NEW CRUNCH WINGS
5 bone-in or 10 boneless...7.50

FUN-YUN CRUNCH WINGS
Sweet and spicy honey mustard wings rolled in our Fun-Yun ring crunch.

WASABI CRUNCH WINGS
Honey soy ginger glazed and rolled in crushed wasabi peas.

THAI ME UP
Thai sweet and Sambal chili glaze rolled in sriracha panko crunch.

These items took a huge market share immediately and alleviated over 25% of the burden off of the traditional wings after just one month.
In our training we focused on multiple elements.

Kitchen Rollout:
Prep and Ordering of the product
Set up of the product
Tasting, Logic and Execution of the product

Front of House Training:
POS Training, order entry and modifications
Review and Tasting
Messaging to the customer
Suggestive Selling of the product

Promoting the product
Signage throughout the restaurant
Social Media and Web images and promotion
Guest Sample Day Preview
Opening day promotions and call to action

When following these rules and the menu engineering tactics layout in this book we have realized far greater success in our menu development impacts ranging from our staff to our guests and certainly to our profits.

While this area of development and planning extends beyond the traditional concept of this book we new we needed to add some these rules as we re often asked beyond the what, that we demonstrate some of the how.

New Item Consideration Delta

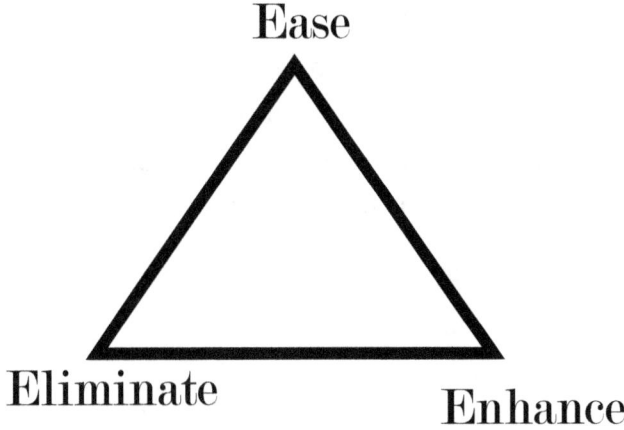

Does it Enhance: the brand, a category, a pricing strategy, gross profit, the season, application or current product selection?

Does it Eliminate: an item, waste, stress in another area, low margin, avoid in the menu?

Does it Ease: the bourdon in another area, the flow, the food cost, labor cost, complication on the menu?

If the item does not meet all of the criteria, use the delta as a consideration or a guide to re-engineer/reimagine the item.

Questions to ask:

- Is there a gap in menu or the market?
- Does this item enhance it?
- Is it unique?
- Is it profitable in % or dollars?
- Will it ease a bourdon?
- Does it eliminate or complicate an item, category or product skus?

Dead Item Reduction Delta

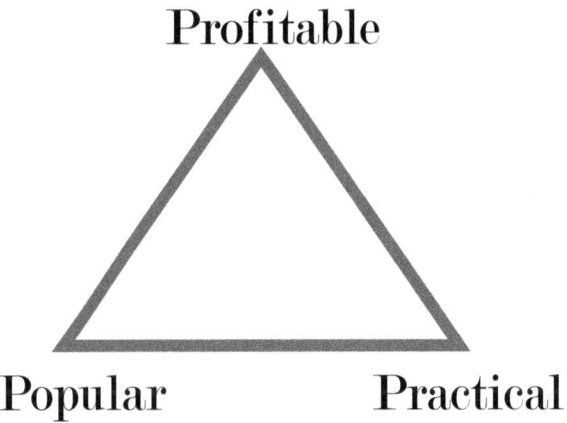

Profitable

Popular   Practical

Is it Profitable: from a food cost or total gross profit dollar perspective?

Is it popular: Inside and out? From the guest perspective, and from an ease of execution or bourdon relief from the staff prospective.

Is it practical: from an application stand point, from a product stand point, from a positioning stand point, a uniqueness stand point.

*If it is underperforming but meets the delta requirements, the first consideration should be to re-engineer/reimagine the item.

Questions to ask:
- Will it create a gap in menu?
- Will it reduce or put a bourdon on skus?
- Is it unique?
- Does it compliment the brand?
- Could it?
- Is it profitable in % or dollars?
- If profitable, but not popular?
- If so, consider a name change, menu relocation, over all flavor profile?
- Will it ease a bourdon?
- Does it eliminate or complicate an item, category or product skus?
- Does it create or ease congestion?

BONUS IDEAS

"Ideas can be life-changing. Sometimes all you need to open the door is just one more good idea."
Jim Rohn

Kids Menus

Make them unique. We are working with a tablet right now that has the menu preloaded as the screen saver and it also has a number of preloaded games and activities that parents approve of. This is a real standout idea and a worthwhile investment to us because we know the power of appealing to kids and keeping them quite, distracted and happy. Parent will pay and stay when you create an environment like this. There are cheap tablets out there that cost less than $100 each and you can lock the content. Our cost to use disposable menus and coloring crayons will pay for this investment in less than a year and with a hell of a lot more impact to the kids and parents alike. We are top of mind with this idea!

Chalk Board/Dry Erase Board

In another concept I consulted with we make a double sided chalk board with a kids menu painted on one side and a free draw platform on the other side. The kids got chalk when they got to the table. The chalk is cheap and unlike crayons you can get chalk out of anything.

We have used a double sided dry erase board in other concepts to create the exact same menu.

Another take on the chalk board idea has been used on slate. Slate was used as the cheese board when the cheese plat comes to the table the type of cheese is written in chalk and the cheese is outlined and listed by number. The sequence of numbers is used to tell the guest which cheesed to eat first so that the cheese is eaten from lightest to strongest. The server educates the guest with this info so that they will enjoy the cheese in the best way possible.

Badass Chalk Idea

Another badass chalk menu idea that I have used works both as a menu concept and as a marketing concept. Right in the path of the front door we will write the evening special on the ground. People are numb to advertising at eye level but writing our special on the concrete below their feet in there walking path jumps out at them. Just like when you are in your neighborhood and you see this done by kids, you always look. The same is true for this application and it gets our special top of mind. We also use it to advertise many other things.

Origami Menu and Origami How To Page

This is another menu idea we have done for kids. The menu is folded up into a cool design and it is interactive. The kids get instructions on the back of the menu page and they get a pack of colored paper to try to fold their own origami art. The same vehicle can be used for adult menus. Use your imagination!

Kids Menus Full Color for Cheap!

Check out www.kidstar.com
This is a site that was created by a restaurateur and I love using their stuff if you want to play is safe.

Clipboards

Clip boards have become a common medium for menus. I used my first clipboard menu in the early 2000's in a joint concept in Costa Rica. We had a gourmet hamburger concept that allowed you to order your custom built burger or you could check a box and pick one of our fixed gourmet burgers. It was a very industrial, affordable way to approach the menu.

In recent years I have used clipboards as an inexpensive platform for a unique menu. Because clipboards are already sized just a bit larger than standard sized 8.5" x 11" paper they are still a good affordable base. We will take the clip off of the board and use the predrilled holes as a place to bind our menu. We will also have them painted or poly coated with a design. These menus allow for a rugged application in an affordable non-traditional way to build a menu.

Industrial

We are always looking for ways to make menus out of industrial suppliers like leather, metal, fasteners, wire, etc. A quick stroll around your local hardware store or big box hardware store will yield some pretty cool results.

Another Chalk board idea

Another idea I have grown fond of is chalk paint. It is perfect for a board menu or any flat surface. The point of the exercise is to get creative and outside of the box on what your menu delivery vehicle will be.

Table Side Menu

There is a great place we love to eat in Seeley Lake, MT. called Lindy's Prime Steak House. This restaurant certainly beckons back to a time forgot in a place even more forgotten but I love this place because of both its simplicity and its flawless execution. There menu is nothing more than a board on a stand that illustrates there 4 menu choices. The server brings the menu over table side shortly after you are seated. It may not work for everybody but it certainly charming and most definitely unique.

The QR Code

A QR code is a quick reference code originally used in auto manufacturing. It is a simple block code generated by a computer program with a simple reader that deciphers the code. There are applications available for any mobile phone to both generate QR codes and read QR codes. In a recent consulting gig I did in Long Beach, CA. I decided to get these really cool QR code stickers made for this concept. The stickers were adhered to the table and a guest could simply scan the QR code to access the menu on their phone. While progress and quite the opposite of simple, it was a great fit for this busy concept with a very young, tech sophisticated drinking crowd. The code is easy to change, the menu is easy to update and the product is very remarkable. This was a supply and demand evolution. This restaurant has a crowd and/or environment that was very rough on menus. While they still had traditional menus available they offered a bonus when you opted for the menu via QR. In the QR code was an embedded discount and when most people were told about this they opted for the QR menu. A little known fact, I have tried and tested, is that QR codes do not have to be just black and white but they can also be read at only 90%. What that means is that I can take up 10% of the surface area and embed a logo or a word... like menu. This is a pretty cool concept that is really, really easy to execute.

Here is the set up. You can create the code just like the one I did below. I did this just while writing this page in about 2 minutes.

The app I used is:
http://qrafter.com

The reader is free. To generate the codes you have to upgrade the app. It cost be $2. Pretty easy and pretty cheap for a unique idea.

On the next page is an example of QR I just created. It will take you directly to my web page. You would basically just paste your link to your menu page on your website and paste it in the link space provided for the QR Generator. You can easily change the colors in any basic computer art program or even in Micro Soft Word by pasting the object in a blank document and changing the color by highlighting the object and going the formatting tab. I added the quick little initial "R" just on this page to prove and test that is would read and it worked like a charm.

QSR QR CODES

QSR or "quick service restaurants" usually have a menu board but often times one of the best consulting pieces of advice I seem to always be giving to concepts like this is to have a menu available before you get in line. The problem with getting in line before seeing the menu is that people usually rush and scan the board menu even faster than a paper menu. When that happens there is much opportunity for revenue lost. I will often post a menu outside on the window or in a frame attached to the building (a great idea for just about any concept) but often I will also plant a QR code throughout the line with a little note that says "See the menu before it's your turn".

MENU BOARDS

As I just alluded to above Quick Service Concepts have a whole different consideration when it comes to menus. The scanning is still there but the pace is much faster when there are people behind you waiting to order and you are prompted to make a faster decision. With that the menu scanning pattern is a little different. Some lines start from left to right and some right to left. People will read or scan the menu in order of the flow of the line. It is with this that you should put your signature items first in line. Then you would list the items you would want sold second and so on. There are plenty of considerations with this type of scanning. My best recommendation would be to move the boards around from time to time and measure the impact that these moves have. This is a little exercise that will yield some very interesting results. You can also play around with the location but for the most part all of the rules laid out in this book will still apply.

Screen-print your menus on your linen napkins. This is an idea I have been playing with for quite a while. It would require an established concept without much menu change to make it logical from a cost standpoint but an idea nonetheless that would be epic. If you decide to use this idea please drop me a line and let me know how it goes. I know it breaks a lot of the rules but it sure as hell will stand out. Ask your questions or get new info and content at http://www.ProfitProPlus.com

Hopefully this book has opened your eyes and your mind to some opportunities to capture more revenue. Go out and implement these ideas and remember if the ideas are better than what you are currently doing, don't wait. Perfect is a hard thing to accomplish and if it is better than what you have now get it going and tweak, modify, measure and repeat… early and often.

ABOUT THE AUTHORS

Thax Turner & Bo Bryant own multiple restaurants and as authors, consultants and serial entrepreneurs they often times find themselves headlong in new business ventures.

As a sales strategist, motivator and speaker Thax; and Bo, the authority in restaurant consulting and brand development, decided to join forces over 5 years ago and together their collaboration has introduced 3 separate books on the restaurant business and a life building strategy book called The Modern Day Spartan.

In there arsenal also sits a Restaurant Software and Technology Company aimed at providing a better back office system for restaurant operators called Profit Pro Plus.

With a wide range of businesses from their marketing company to their radio show, their restaurants and consulting company or the multiple joint ventures and project development concepts, Thax and Bo are constantly working to improve the lives of the Entrepreneur and their businesses.

Tune into The Thax & Bo Show: Entrepreneur

Radio every week through itunes or catch them on TheThaxandBoShow.com to listen to previous shows and get caught up to speed on all the cutting edge concepts that are proving to be the most impactful and elegant solutions to small business owners around the world.

PROFIT PRO

www.ingramcontent.com/pod-product-compliance
Lightning Source LLC
Chambersburg PA
CBHW051918170526
45168CB00001B/442